Louisiana Life Series, No. 8

WHITE GOLD

A BRIEF HISTORY OF THE LOUISIANA SUGAR INDUSTRY 1795 - 1995

by

Glenn R. Conrad
and
Ray F. Lucas

Published by
The Center for Louisiana Studies
University of Southwestern Louisiana
Lafayette, Louisiana

Library of Congress Catalog Number: 94-074142
ISBN Number: 940984-92-X

University of Southwestern Louisiana
Lafayette, Louisiana

Contents

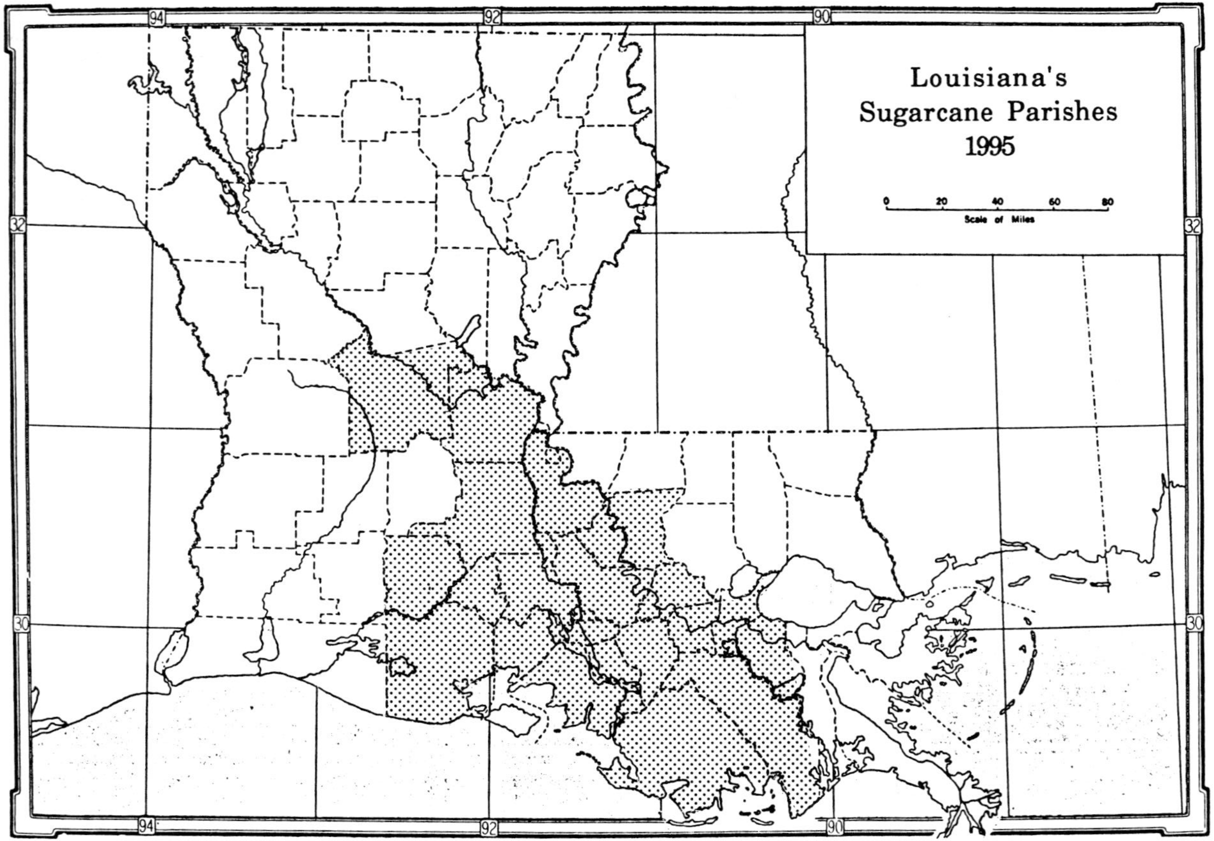
Louisiana's
Sugarcane Parishes
1995
0 20 40 60 80
Scale of Miles
94
92
90
32
30

Introduction

In 1979, during the height of the oil boom in Louisiana, word reached the Center for Louisiana Studies that the state's sugar industry was fast fading out after being part of the state's economy for nearly two hundred years. Farmers and would-be farmers in Louisiana's sugar belt, it was said, were deserting the soil for high-paying jobs in the oil industry. It was obvious that vast acres of farmland lay idle and that the oil industry and its related service industries were moving in where undulating sugarcane once grew.

In response to the suggestion that the days of the state's sugar industry were numbered, the staff of the Center organized a project to document through words, photographs, and other images the evolution of the industry since the commercial production of sugar began in 1795. The completed project, entitled *Green Fields: Two Hundred Years of Louisiana Sugar*, resulted in a 25-panel photographic exhibition accompanied by a catalog telling the story of sugar in Louisiana.

From the beginning of the project, however, the Center staff quickly realized that reports of the imminent demise of the state's sugar industry were greatly exaggerated, if not, indeed, contrary to fact. Research for the project clearly indicated that the industry was on the verge of a technological breakthrough that would revolutionize sugar production in Louisiana from cultivation to manufacturing.

In the years that have passed since the Center's project, that technology has come to be applied to the industry, and new technology promises exciting changes as sugar production enters its third century in Louisiana. As we come to the bicentennial year, the Louisiana sugar industry is more vibrant than ever.

It is the purpose of this brief history to review the course of the industry during the past two hundred years and to examine the technological advances and human determination that have brought it to the dawn of the twenty-first century.

Glenn R. Conrad and Ray F. Lucas

I

Early Louisiana and the Birth of an Industry

In 1682. after receiving the commission of King Louis XIV of France, René-Robert Cavelier, sieur de La Salle, led an exploratory expedition from New France (Canada) to Louisiana. The expedition planned to retrace the route taken earlier by the explorers Louis Joliet and Father Jacques Marquette. The La Salle party traveled across the Great Lakes, then down the Wisconsin River to the Mississippi River. Continuing down the Mississippi, passing the present site of New Orleans, the exploratory party eventually made camp near the mouth of the great river. There, La Salle claimed the Mississippi watershed, the heart of North America, for France, and he named the new French territory Louisiana, in honor of Louis XIV. The expedition then returned to Canada, following the route that brought it to Louisiana.[1]

La Salle's activities were largely owing to the fact that during the seventeenth century many nations of Western Europe had launched rival expeditions to found colonies in the New World. These nations were motivated to do this by expectations of huge economic benefits the colonies could bring to the mother country. Because of this rivalry, nations claiming territory needed to establish colonies or some type of military outpost to secure their claim to a specific part of North America. For this reason La Salle had been commissioned to establish the French presence on the Mississippi River at a point near the gulf coast. In 1684, La Salle again set out for Louisiana, this time with a few settlers to establish a colony. Owing to navigational error, however, the expedition passed the mouth of the Mississippi River and landed in Southeast Texas where the nascent colony was established. The colony endured many hardships, including the loss of its leader. Shortly after the death of La Salle, the Spanish government learned of the small French colony that had unknowingly been set up on land claimed by Spain. Spanish authorities dispatched troops to find the colonists and expel them. Mexican troops eventually discovered the French post and took into custody the few surviving settlers.[2]

Because of La Salle's failure to establish an enduring colony and lack of money in the royal treasury, the French government did not again attempt to colonize Louisiana for several years, not until 1698. Pierre Le Moyne, sieur d'Iberville, and his brother Jean Baptiste Le Moyne, sieur de Bienville, led the expedition. The Le Moyne brothers were instructed to establish a military outpost near the mouth of the Mississippi, just as La Salle had been instructed earlier. The outpost would have as its purpose safeguarding the entrance to the Mississippi River, thereby protecting the backdoor of New France from European rivals, especially the English.

In 1699, the Le Moyne expedition landed on the coast of the present-day state of Mississippi. A small settlement was established in the area of present-day Biloxi and was named Fort Maurepas. Throughout the next twenty years, several small settlements were established, stretching from Mobile on the bay of that name to Natchitoches on the Red River.[3] In 1718, Bienville founded New Orleans on a ridge that extended between Lake Pontchartrain and the Mississippi River at a point about ninety miles upriver from the Gulf of Mexico.[4]

The French military outpost of Louisiana grew slowly, and while there was some agricultural production, colonists could not always produce enough food to feed themselves. A major factor for the lack of food production was that Louisiana had not yet been colonized by farmers and artisans. Moreover, during the early years of the colony many individuals sought quick wealth through treasure hunting or fur trapping and trading.[5] Then, in the early 1720s a policy shift transformed the military nature of the Louisiana outpost into a settlement colony. Now, farmers and artisans were recurited to emigrate to Louisiana. At the same time crops that could hopefully make a profit were introduced into Louisiana. The profit motive and the semi-tropical climate of the Gulf Coast were probably the main reasons for the introduction of sugarcane into the colony. Sugarcane had been grown widely and profitably in the French and Spanish island colonies of the Caribbean Sea. Cultivation of sugarcane was profitable because world demand for sugar was growing quickly during the eighteenth century.

The introduction of sugarcane into Louisiana should probably be credited to Iberville during the opening years of the colony. Thereafter, small amounts of cane were grown, but most attempts at cultivation ended in failure. One of the more encouraging of these attempts was that of the Jesuit fathers who, in the 1750s grew sugarcane on their plantation then located on what is the present-day site of New Orleans.[6] Any sugar or syrup produced from the Jesuits' crop was probably used only for their own consumption. Sugarcane was grown and sugar was produced commercially in Louisiana as early as the late 1750s. Joseph Villars Dubreuil, a militia captain, operated a sugar mill in New Orleans in 1756. The product of his early milling attempts was a liquidity raw sugar. The full granulation process remained a mystery to Dubreuil.[7] He died in 1757, and his mill was sold to a person named Mazan. In 1763, Mazan sold one half of his sugar operation to Louis Boré.[8]

In the meantime, however, the political map of North America was being significantly altered. In 1754, France and Britain began the final contest for imperial control of the continent. War broke out in the colonies and quickly spread to Europe. The war is known as the Seven Years' War in Europe and the French and Indian War in North America. By 1757 Britain was near defeat, but all that was to change in the next year. In 1758 British forces laid siege to the French fortress of Louisbourg on Cape Breton Island and were astonished at how quickly the defending forces surrendered, paving the way for the British strike at the very heart of New France, the citadel of Quebec.

In 1759 the siege of Quebec began, but, being that it was late in the year, the British knew that they had to move swiftly to capture the city or winter weather would force them to lift the siege. In an event remarkable in military annals, the British enticed French forces to leave the fortified city and engage in battle on a plain a few miles distant. In the ensuing battle a large part of the French army was defeated; however, French troops still occupied the city of Quebec. The British laid siege, and the city surrendered in September 1759. One year later, British forces attacked and captured the last major French stronghold, Montreal. These three crushing defeats caused France to abandon Canada; however, Louisiana remained a French possession.

Near the end of the war, in 1762, France secretly transferred Louisiana to Spain by the Treaty of Fontainebleau. With the loss of Canada, the French were certain that the British would not be satisfied until they had ended all French influence on the continent. Thus, admitting defeat, France transferred Louisiana to Spain rather than see it fall into British hands by default.[9]

Louisiana thus became a Spanish colony. French residents of Louisiana were distressed by the transfer and launched several unsuccessful appeals to the French government to maintain control of the province. As the years passed, Louisianians remained unhappy with Spanish rule, and as a result, in 1768, some residents of New Orleans and environs rebelled against Spanish control. A year later, the Spanish government dispatched General Alejandro O'Reilly to put down the rebellion and firmly establish Hispanic control over the colony.[10]

Their authority secured, the Spaniards sought to make Louisiana profitable. They moved to streamline government, improve communications and transportation, promote the cultivation of food stuffs, and encourage the production of cash crops. Among the last were indigo, flax, hemp, cotton, sugar, and tobacco; crops, however, not necessarily suited for the semitropical climate of southeastern Louisiana.[11] In the case of sugarcane, the cession of Louisiana to Spain, the rebellion, and years of severe weather serve as plausible reasons for the decline of sugarcane cultivation between 1763 and 1790.[12] At the same time Spanish authorities were encouraging planters to undertake large-scale cultivation of indigo as a cash crop. One planter to respond favorably was Etienne Boré, the son of Louis Boré who had purchased half of the Mazan sugar plantation and mill in 1763. A few Louisiana planters continued to experiment with sugarcane cultivation, and, though most of these early attempts ended in failure, the experiments taught the planters valuable lessons.

For the moment, however, most planters were content to grow and process indigo. But, as the years passed, indigo cultivation became increasingly less profitable because: first, indigo cultivation was at best marginal in the climate of southeastern Louisiana; secondly, there was increasing competition from South American and Atlantic East Coast colonies, resulting in the depression of prices for Louisiana indigo; thirdly, Louisiana experienced an invasion of pests that virtually destroyed the indigo crops of 1793 and 1794.[13] These problems and others forced planters to attempt cultivation of other crops. Etienne Boré turned to sugarcane.

The accomplishment for which Boré is best remembered is his successful sugar crop of 1795. Contrary to a popular misconception, Boré was not the first person to granulate sugar in Louisiana or anywhere else. Sugar was being granulated in the Middle East as early as 1100 A.D., and in Venice by the fourteenth century. Also, as noted above, Dubreuil did granulate, however imperfectly, some of his sugar crop during the late 1750s, about thirty-five years before Boré.[14]

With the indigo situation what it was, by the 1790s some Louisiana planters began experimenting once again with sugarcane cultivation. In this process they learned the absolute need for good drainage of sugarcane fields, which are usually located in that part of the state which is between five and fifty feet above sea level. They developed the practice of windrowing to counter the freezing temperatures which sometimes occurred in Louisiana during harvest season and which would damage or kill sugarcane. Windrowing was the process of cutting as much cane as possible during a period of freezing temperatures, laying it in the furrows, and placing a layer of dirt or bagasse (the residue of cane after the juice has been removed) over the cut sugarcane until it could be milled. Early planter experiments, however, proved that windrowing was only a short-term resolution of the freezing problem because sugarcane, once cut, cannot be stored for periods of time without deterioration of the sugar content. Finally, planters came to realize that they should grow only that amount of cane which could be processed in mills before the usual freezing temperatures of late December and January killed the cane crop.[15]

Most experiments with milling cane and producing raw sugar (a granulated brown sugar heavy with molasses) were not successful in the 1790s. Joseph Solis, a pioneer of sugarcane cultivation in Louisiana, might be regarded as the link between the experimental sugarcane industry of pre-1795 and the new era that was about to begin. Solis had planted sugarcane and from it produced molasses which he distilled to produce a liquor called taffia. Apparently, however, this enterprise was not profitable because in 1794 Solis sold his land, sugarcane, and mill to Antonio Mendez. Mendez and other planters were hiring refugees from Saint-Domingue (Haiti) who had experience as sugarmakers. A sugarmaker was responsible for boiling sugarcane juice to a point where, through the injection of a catalyst, the granulation process would begin. The sugarmaker hired by Antonio Mendez during the 1794 grinding season was Antoine Morin.

Mendez was successful in producing sugar from his 1794 crop, but his venture resulted in only a small profit. His lasting contribution to the future of the Louisiana sugar industry was that he proved that there were varieties of sugarcane which would sufficiently mature during Louisiana's relatively short growing season to possess a high sugar content and therefore granulate, something previously used cane varieties would not do. In 1795, Mendez sold his entire crop to other planters for seed cane. Etienne Boré not only purchased and planted some of the Mendez cane, but he also hired Morin to be his sugarmaker for the 1795 season. Boré's actions in this regard proved to be successful. The crop produced 100,000 pounds of sugar that resulted in a profit equivalent to $5,000. Boré's next crop, in 1796, was also a success, and with this accomplishment the modern Louisiana sugar industry was launched. For this reason Etienne Boré has earned widespread recognition as the "founder of the modern Louisiana sugar industry,"[16] but Antoine Morin, the experienced sugarmaker from Haiti, deserves similar recognition.

Etienne Boré and other planters now began the long task of addressing all the problems associated with cultivating sugarcane in Louisiana's unusual growing conditions. The sugarcane growing region of Louisiana has a semitropical climate. Each year the spring, summer, and early fall are warm with an adequate amount of rainfall (although late spring and summer droughts or "deluges" can seriously damage the cane crop). Louisiana's winter weather, however, is unpredictable. Some years the winter is quite mild with perhaps frost but no freezing conditions. Some years freezing temperatures occur only in January or February, after the sugarcane harvest. In still other years there can be periods of several days of below freezing temperatures. Approximately every decade or so, winter will bring snow to Louisiana's sugar-producing region.

The Louisiana sugarcane growing season extends over approximately nine months from mid-March to late December. Sugarcane in tropical regions usually needs twelve to fourteen months to reach full maturity. Thus, Louisiana sugarcane is immature when harvested, that is to say before being fully grown. Nevertheless, during its growing season, Louisiana cane usually builds up a sucrose content sufficient enough to make production of sugar profitable. Louisiana's sugarcane growing region is mostly confined to the area of the delta plain, but sugar can be grown on the prairie terraces to the east and west of the delta plain. The soil is fertile and contains the minerals and nutrients needed for sugarcane growth. The general topography, however, is flat; therefore, to achieve the necessary drainage planters have long devised ways to grow sugarcane on unusually large rows, generally spanning six feet from furrow to forrow. So, Louisiana is perfectly suited for growing sugarcane if late fall and early winter weather does not produce freezing temperatures and late spring or summer weather does not bring with it prolonged drought or excessive rainfall.[17]

News of Boré's successes spread rapidly across southeastern Louisiana. Many planters sought seed cane for planting. In the next five or six years more and more plantations were given over to sugarcane cultivation. In 1797, two years after Boré's initial success, 550,000 pounds of sugar were exported from New Orleans. By 1801 Louisiana had seventy-five sugar mills producing 5,000,000 pounds of sugar. That year, sugar sold for eight cents per pound. That price translated into a gross return of $400,000 for Louisiana sugar planters, not taking into consideration the sale of molasses and taffia.[18] Agricultural pursuits were rapidly changing in Louisiana, but the realm of Louisiana agriculture was not the only place where rapid change was occurring. Events on the international scene were also in flux and would eventually affect the colony.

In 1800, Napoléon Bonaparte intimidated the government of Spain into retroceding Louisiana to France. The transfer became official on October 1, 1800, with the signing of the Treaty of San Ildefonso; however, the transfer was kept secret because Bonaparte could not then take control of Louisiana. Quickly changing events in Europe would serve to alter further the destiny of Louisiana. Bonaparte's wars of aggression were costly. He desperately needed money to pay for his various schemes of aggrandizement. Moreover, the troops he planned to use for the occupation of Louisiana had to remain in France.

As events unfolded, rumors of the secret transfer of Louisiana to France spread to Washington, D. C., and the White House. The rumors, if based in fact, were of great concern to President Thomas Jefferson. If France controlled Louisiana, that is to say mid-America, Bonaparte might be tempted to blackmail the United States for use of the lower Mississippi River and the port of New Orleans. The Mississippi, with its outlet to the gulf, was vital to the economic growth and political union of the expanding United States. As a result of these concerns, Jefferson instructed the United States ambassador to France, Robert R. Livingston, to offer to buy the Isle of Orleans from the French. The Isle of Orleans was considered to be the land east of the Mississippi River and south of Lakes Maurepas and Pontchartrain. The city of New Orleans was located on the Isle. Through the purchase of this strategic territory with its city, the United States would assure itself of the free flow of commerce to and from the Mid-West.

James Monroe soon joined Livingston in the negotiations. At first the French seemed opposed to a sale. Then, unexpectedly, the French foreign minister offered the Americans the entire province of Louisiana. The United States representatives did not know how to react. They were only authorized to buy the Isle of Orleans; nevertheless, Monroe and Livingston took it upon themselves to buy all of Louisiana. After several days of price negotiations, the French agreed to sell Louisiana for about $15,000,000. On May 2, 1803, with the signing of the purchase agreement, the United States almost doubled in size. Events now moved swiftly. On November 30, 1803, the colony of Louisiana was officially transferred from Spain to France. Roughly three weeks later, on December 20, the French transferred Louisiana to the United States.[19]

The Louisiana Purchase would be fraught with consequences, among which one of the more important was that the vast trans-Mississippi region was now open to Anglo-American settlers. The Spaniards had encouraged Anglo-American immigration in the 1790s, but most of these immigrants had settled in the area today known as the Florida Parishes. Once Louisiana became part of the United States, Americans began to move into all parts of the territory, including the sugar producing region of the Mississippi delta. The newcomers would join the older inhabitants of the region in the further expansion of the sugar industry.[20]

Sugarcane cultivation spread across the southern part of the Territory of Orleans (the present-day state of Louisiana) after 1806. In that year sugarcane was being grown along the Mississippi River from Baton Rouge to below New Orleans. Soon sugarcane made its appearance on the banks of Bayous Lafourche and Terrebonne, and before long its cultivation spread to the Teche country. There are several explanations for the rapid spread of sugar production in Louisiana. One was the influx of American settlers, many of whom carried with them ideas of improved farming techniques. Another explanation can be found in the abundant supply of cheap labor—slave labor. As the sugar plantations of Louisiana developed, slaveholders in the Middle Atlantic and southeastern states found a ready market for their excess slave populations. Thus began the woeful business of "being sold down the river." A third factor for the rapid spread of the sugar industry in Louisiana at the beginning of the nineteenth century was the arrival of thousands of refugees from the revolution in Haiti. Many of these people brought with them experiences in the sugar industry which could be easily applied to conditions in their new homeland. Finally, a factor of undetermined significance was the decline of the European beet sugar industry in the face of the Napoleonic wars.[21]

When the United States bought Louisiana, the area of the present state, then called the Territory of Orleans, had a population large enough for immediate entry into the Union. Louisiana, however, would not gain statehood until 1812. The federal government presumed that the *ancienne population*, the Creoles, needed a period of adjustment to become part of a republican system. Whether or not that was the case remains academic. In 1812 Louisiana entered the Union—the Creoles were fully republicanized but they remained far from being Americanized.

The same year that Louisiana became a state, the United States declared war on England. The War of 1812 did not really affect Louisiana until the end of 1814 and beginning of 1815. Until that time Louisianians regarded the war as being something occurring on the distant East Coast. In late 1814, however, a British fleet was sighted in the Gulf of Mexico. General Andrew Jackson believed the British were headed for New Orleans in an attempt to penetrate the Mississippi Valley. Jackson headed for the lower Mississippi with several hundred troops, and more from Tennessee and Kentucky joined him later. Louisianians in great numbers volunteered to help defend their homeland from the English invasion.

The British forces came ashore just south of New Orleans. Jackson and his ragtag army had set up a defensive line between the Mississippi River and a cypress swamp to the east. The battle was soon coming and lasted from December 23, 1814, through January 8, 1815. The main and final assault occurring on the latter date. The British launched a massive attack on the

American defensive line. The Americans, safely behind their makeshift defenses, decimated the experienced British troops. The British commanding general, Sir Alexander Pakenham, died as a result of injuries received during the battle. New Orleans was saved. Several days later the British boarded their ships and left Louisiana.

The Battle of New Orleans was an ironic event. It occurred after the warring parties had ceased hostilities. American and British diplomats agreed to peace, as set out in the Treaty of Ghent, on December 24, 1814. No one in Louisiana, British or American, had any way of knowing that the war had ended, and so ensued the battle of two weeks later. Louisiana was safe, and Andrew Jackson was a national hero.[22]

II

Louisiana's Antebellum Sugar Industry

Louisiana's admission to the Union greatly enlarged the market for the state's sugar crop. In response to this, the state's planters increased the number of acres in sugarcane production and thus increased the amount of raw sugar produced. By the early nineteenth century the national demand for sugar was so great that Louisiana could not satisfy that demand. The relative shortage caused the price of sugar to reach the unprecedented price of eighteen cents per pound in 1816. Such a return on their investment encouraged planters to expand sugarcane cultivation even more. The protective tariff of the United States abetted the growth of the industry. When Louisiana entered the Union in 1812, the tariff on raw sugar was 2.5 cents per pound. During the War of 1812, the tariff was as high as 5 cents, but was lowered to 3 cents in 1816.[23]

With many signs pointing to "good times" in the industry, sugarcane cultivation spread from the Mississippi River westward into the Teche and Vermilion watersheds and northward along the Mississippi River above Baton Rouge, into Pointe Coupee Parish and eventually into the Red River Valley.[24] Meanwhile, news of the sugarcane planters' prosperity spread like wildfire across the United States. Farmers elsewhere now looked with envy on the situation in the Louisiana cane fields. Magazine articles and reports given by travelers to the area depicted the sugar region of Louisiana to be a planter's paradise. The prospects of cheap land and good soil encouraged many planters from Virginia to Misssissippi to relocate in Louisiana for a chance to strike it rich with the region's "white gold."[25]

One large group of agriculturalists to migrate to Louisiana were tobacco farmers from the upper South. Tobacco had been the main commercial crop of Virginia and North Carolina since colonial times and continued to be so well into the nineteenth century. Unfortunately for the East Coast planters, tobacco cultivation proved to be excessively harsh on soil nutrients. Thus, by the 1820s, soil conditions in the tobacco-growing region were being described as "burned out." As a result, many tobacco planters began moving westward to virgin lands in Kentucky and Tennessee. Other tobacco growers were of the opinion that tobacco was no longer worth the exhaustive effort and were becoming attentive to announcements of the periodic sale of public lands in south Louisiana. These lands were said to be extremely fertile and could be bought for about $1.25 per acre. This further encouraged many tobacco and other cash-crop planters to sell off their exhausted lands for whatever they could get, pack up their families and slaves, and move to Louisiana. Many of these Anglo-American planters settled in Louisiana's sugar region, purchasing large amounts of land and equipment to begin their experience with growing sugarcane. These newcomers and the older Creole planters gave birth to what Hollywood and romantic novelists have portrayed as the golden age of Louisiana's sugar industry.[26]

The most fertile and best drained lands of Louisiana's deltaic plain lie adjacent to the rivers and bayous of the region. The soil is alluvial in nature, that is, it has been deposited in layers whenever the rivers and bayous overflowed their banks throughout eons past. The region's rivers and bayous are thus flanked by a natural levee, usually along both banks. The natural levees follow the waterways from their source to where the stream enters another or empties into the Gulf of Mexico. The cane-growing area of the deltaic plain is on land ranging from 5 feet to 50 feet above sea level. More often than not, the highest land, the best drained land, is found on the natural levees. From these levees the land slopes downward away from the stream until it reaches a swamp or marsh, usually called backland. Thus, the natural drainage of the region often is not into a nearby stream, but away from it as rainwater and floodwater runs down the natural levee to the backland.[27]

The configuration of the land in Louisiana's sugar region required planters to learn early the value of drainage and levee protection. Among the first things a planter did when developing his antebellum plantation was to reinforce the natural levee along his waterfront (In Louisiana before the Civil War almost all plantations fronted on a navigable stream.). To do this a planter piled dirt on the natural levee until the levee at water's edge was higher than the average annual high water mark of the waterway. Planters were keenly concerned about the maintenance of their levees because the consequences of a break and the ensuing flood could spell economic ruin. But a planter's concern about levee protection went beyond his own precautions. If a neighbor did not properly maintain his levees and a break occurred, the entire neighborhood could be flooded. Moreover, the planters had to be ever on guard of being "stabbed in the back." Flood waters could rise from the backland swamps and marshes after heavy rainfalls (The average annual rainfall in south Louisiana is about 60 inches, but rainfall of 100 inches is not uncommon), even though the levees along the waterways were in good order.[28] Many planters addressed the problem of backland flooding by building a levee across the back of the cane fields.

The planter, at the same time, had to maintain a constant check on drainage within cane fields. Adequate drainage is important because sugarcane while thriving in damp soil cannot long survive wet soil conditions. Cane may survive flooded conditions for a very short period of time, but roots begin to rot if the water is not soon drained off. In response to sluggish drainage usually found in the cane-growing region, planters had ditches dug, ordinarily draining water from the highest part of the natural levee to the backlands. These drainage ditches, sometimes actually canals, varied in width and depth according to the planter's needs. There were also smaller ditches running parallel to and perpendicular to the larger drainage ditches and canals to conduct water from the interior of the field to these larger conduits. After the introduction of steam power, many planters installed steam water pumps to pump water over the protection levees into the back lowlands. Down to the present, drainage is a key factor in cultivating sugarcane, and the drainage patterns of the past, now vastly improved, remain the drainage systems of today.[29]

With his drainage system developed, the antebellum planter was ready to plant sugarcane. The soil of the fields was broken with crude plows pulled primarily by oxen, but horses and mules were sometimes used. Furrows were made in the plowed soil with a width of approximately six feet between them. Once this was done the process began of forming the row by repeated passes of the plow to throw the dirt into a bank or row. What was produced was an unusually large row, both in height and width, running the length of the field, or in subdivisions of the field delineated by unplowed strips called "headlands." The large rows were all part of the field's drainage system, and they can still be seen in today's sugarcane cultivation.

Once the rows were formed and a few good showers of rain had packed the soil and stabilized the row, a small forrow was plowed open on the top center of the row. It would be into this opening that the seed cane would be dropped and covered with soil and tapped down to prevent the soil from washing away from the cane during the next rainfall. Sugarcane seed is actually stalks of mature or near-mature cane. Each year sugarcane is harvested for two purposes: planting a portion of next year's crop; and, the bulk of the harvest, for processing into sugar.

Planting cane, until recent times, was an especially long and arduous process because it had to be done almost entirely by humans. Until mechanization in the mid- to late twentieth century, mule and oxen drawn carts full of seed cane were pulled through the fields (the axles of these carts being the exact same width as the furrows flanking a row). Field workers would then take cane from the cart and place the stalks, end to end, in the previously prepared planting bed atop the row. Planting cane was a skill of sorts and could not be done haphazardly because the sprouted cane needed to be straight down the middle of the row for cultivation purposes. Workers, using hoes, would then cover the seed cane with two to four inches of soil.

For a long time, cane was planted either in the spring after grinding or in the fall before the harvest began. Today, most cane, if not all, is planted in late August, September, and early October. Planting can be extended into November if a wet fall season delays farmers from getting into their fields. With cane planted in the warm days of late summer and early fall, the plant may sprout and grow to some considerable height before being cut down by a December or January freeze. The foliage thus dies, but the plant's root system remains safe below ground level to sprout again when warm spring weather arrives.[30]

As the cane begins to grow in springtime so do a host of weeds, and frequent cultivation becomes necessary. In antebellum days a brigade of workers, hoes in hand, would attack the grass and weeds to keep them from smothering the tender cane shoots. Cultivation is as important to sugarcane as is proper drainage. While cultivation progressed, drainage maintenance continued. Today, the gangs of field hands are gone as cultivation and drainage work are done mechanically. Weed control has been accomplished through the use of various herbicides.

Cultivation in the past, as is the case today, continued until the stalk and foliage of the cane were tall enough to block sunlight needed by the weeds to grow. Usually by mid-June or early July, sugarcane is tall enough to survive on its own. The time between the end of cultivation and the beginning of planting in late August is known as "lay-by." In antebellum times this was not a period of rest. Other crops had to be harvested, and the plantation buildings and equipment repaired and maintained for the upcoming harvest and grinding season. It was a time when workers would go into the swamps to cut wood (they also cut wood during the slack season between the end of the harvest and beginning of cultivation) for the fires that boiled the sugarcane juice in the process of making sugar.

When August and September rolled around planting began again. The planter did not have to replant all of his fields each year because sugarcane is a ratoon crop. When harvested, whether by hand as in the past, or by mechanical canecutters today, the stalks of cane are cut at ground level and the roots of the plant are left in the ground. The next spring a new plant sprouts from the roots. Usually, three crops of cane can be harvested off of one planting with, under normal circumstances, diminishing returns on the second- and third-year productions. After the third harvest, in the past, the roots were plowed up and the field was left fallow to replenish nutrients in the soil before another crop was planted. Returning nutrients to the soil was achieved rapidly by planting legumes after cane harvest and plowing them into the soil before the next cane planting.[31] Today, with commercial fertilizers, many cane farmers do not see the need for the fallow year.

Harvesting, before the arrival of the mechanical canecutter, was performed by large numbers of workers wielding long knives. The worker would clutch a stalk of cane and begin an unending motion that would cut the stalk at the ground, sweep up and trim off the unripe joints at the top of the stalk, and then slash downward removing dead foliage still clinging to the stalk. The cut and stripped stalk would then be thrown into a pile to be loaded by other workers into a cart for transfer to the mill. Often, the cane cut was thrown directly into a cart which when loaded moved off toward the mill.

Transportation, however, varied in time and in place. On some plantations, the cane was loaded onto barges and floated down the drainage canals to the sugarhouse. In the early nineteenth century some plantations had railroad tracks laid throughout the fields on which mule drawn carts would bring the cane to the sugarhouse. Later in that century and into the twentieth century, some plantations used steam-powered miniature locomotives to pull cars loaded with cane from the fields to the sugarhouse. These plantation locomotives were generally known as "dummy locomotives" and the narrow-guage plantation railroads referred to as "dummy lines."

In the nineteenth century, loading cane into whatever type of conveyance to take it to a mill was a backbreaking task usually accomplished with human and animal muscle. When, however, steam power was introduced, planters began to design derricks which, using grabs or slings, could lift the cane from the field cart and place it onto the conveyance which would take it to the mill. At the mill, another similar, but sometimes larger, derrick would remove the cane from the conveyance and position it for the milling process. With the invention of the combustion engine, most plantation derricks were adapted to this energy source. For a long time, the wooden or metal jackleg plantation derrick was a common sight in cane country. In the middle of the twentieth century, with the coming of the mechanical cutter, also came the mechanical cane loader which moves through the field pushing cut cane, which is laying across two rows, perpendicular to the oncoming loader, called the "heap row." When the bundle of cane is the right size, a mechanical arm with a grab attached to it moves down from the loader, raises the cane from the ground, and deposits it in a tractor-driven cart which runs alongside the loader. In the late twentieth century this technology has been perfected to the point where, in dry-field conditions, the forward movement of the loader and accompanying cart are constant. It is more than likely that before the dawn of the twentieth-first century, technology will have moved on to produce a sugarcane "combine" which will cut the cane, remove the foliage, segment the stalk, and deposit the cane in an accompanying cart for transfer to the mill. Though the jackleg derricks are now a thing of the past, the few still to be seen in cane country are reminders of the ongoing technological advancements in the cultivation, harvesting, and transportation of sugarcane.

Once sugarcane is cut, it needs to be transported to the mill as soon as possible because deterioration of sucrose content begins. Until mechanization of the industry, the sugar mill could take cane for processing almost as fast as it could be cut in the fields. Sometimes the mills were faster than the cutters. It is a well recorded fact that when this happened in antebellum times, slaves cut cane all day and well into the night, working by the light of the moon.[32] Today, the harvest operation has to be slowed through the use of daily delivery quotas so that the mills can be assured of always grinding freshly cut cane. Until mechanization of field operations, there were all sorts of problems to confront farmers between field and mill. Any farmer of the past and many at present can relate tales of the harvest caught up in weather conditions that caused farm equipment to become bogged in the fields, thus delaying the harvest for days, sometimes more than a week. The tales of canecarts bogged to the axles are legion, so too those of tractors wallowing helplessly in the muck of a rainy November.

After the coming of the central sugar factory, about the turn of the century, and down to mid-century, most farmers in the Louisiana cane belt found their operation of transporting cane from the field to the factory hampered by substandard roads and bridges. After World War II local and state governments began hard surfacing rural roads and building substantial bridges. Thus, the problems of the farmer in transporting cane to the mill have been significantly reduced. As the century closes, cane is being transported by diesel-powered trucks pulling trailers, holding forty tons of cane, over hard-surfaced highways, some of them four-laned. In some instances the distance between field and factory is seventy miles one way.

With the sugarcane cut and delivered to the sugarhouse, the milling, or grinding, process begins. To many Louisianians of the cane region, this time of the year is simply referred to as "grinding." French-speaking Louisianians refer to it as the "roulaisant." Whatever the label, it was and remains a time of great activity in and around sugar mills.[33] Traditionally, sugar mills have operated nonstop until the harvested crop is processed, a period of about two-and-one-half to three months.

Early sugarhouses were little more than simple sheds located on the growers' plantations. Under the shed were the various apparatuses for crushing the cane, extracting the juice, boiling it, making sugar, and loading it into wooden barrels. From this primitive beginning in about two thousand plantation sugarhouses, the manufacturing process has developed to the point where, in late twentieth century, twenty mills, using the latest technology, grind some of the largest sugar crops produced in the state.

The various steps in transforming sugarcane juice into raw sugar have remained much the same through the two centuries of sugar production in Louisiana. In the early days of manufacturing, the cut stalks of cane were hand-fed into a set of rollers, one atop the other, designed to crush the cane and force out its juice. The rollers were turned by animal power, but as time passed, steam, diesel, and natural gas engines were applied to the rollers. The original rollers were made of wood and did not last very long. Then steel bands were applied to the wooden rollers to achieve longer use. Finally all-metal rollers were installed.

The juice from the crushed cane was collected in vats below the rollers (often the cane was passed through another set of rollers to press out more juice if possible). When the process of "grinding" the cane was completed there remained a fibrous residue called bagasse. In the early years of sugar production, bagasse was considered to be a waste product. Then, a few planters began to use it as fertilizer or for windrowing cane. As the years passed, many uses were found for bagasse, including use as fuel to fire the sugar mill boilers.[34]

During the antebellum era, sugarcane juice collected in the vats beneath the rollers was next filtered to remove pulp and foreign matter that may have dropped from the cane during the crushing process. After being filtered the juice was conducted to cast-iron kettles for boiling. There were four open-air kettles, each one having its own name and size. They were the "grande," the "flambeau," the "syrup," and the "battery." These kettles were aligned from the largest to the smallest.

The juice would stay in the grande until it boiled. The boiling process was to evaporate water in the cane juice and to kill any impurities in the liquid. The impurities would come to the top of the boiling juice and workers would skim the foamy substances with long-handled ladles. Once this process was completed, the juice was ladled into the flambeau where the same process would be repeated, and repeated a third time as the thickening juice was ladled into the syrup kettle. In the flambeau and syrup kettles lime would be added to extract remaining impurities. The final step in boiling the juice, now syrup, was to take place in the battery.[35]

The sugarmaker, a well-experienced person, would oversee the syrup boiling in the battery. When it had reached the right temperature and the proper consistency, the sugarmaker would attempt a "strike." The strike occurred when the boiling mass began to produce sugar crystals, indicating granulation was near. At just the right moment, the sugarmaker ladled the syrup into vats to be cooled. If the strike had occurred at the right time the syrup would begin to crystallize in the vat. The process would continue until the cooling vat was full. If the strike did not occur at the right time, the syrup would cool in the vat and become a mass of molasses. Thus, the greatest sugar crop in the field could be rendered almost worthless if the juice from the cane was not properly processed. The sugarmaker was obviously one of the most important people on the plantation.[36]

After cooling in the vat, the raw sugar, heavy with molasses, was taken from the vat and placed in hogsheads. A hogshead was a large wooden cask used for two purposes. First, it was the container for shipping sugar from the sugarhouse to market. Secondly, and perhaps of greater importance, the hogshead served as agent to remove molasses from the sugar. Once the molasses-heavy sugar was packed into the hogshead, holes were drilled in it to permit the molasses to drain through the sugar and out of the cask, leaving, under ideal conditions, a fair grade of raw sugar in the hogshead.

This process of draining molasses from the hogshead went on while the casks were still on the plantation waiting to be shipped to market or, if the sugar was sold immediately, the draining process took place on the vessel taking the sugar to market. The molasses usually drained into a receptacle and was later used for other purposes or marketed. The draining of molasses in this fashion, however, could be tricky business. Many factors affected the condition of the sugar and molasses in the hogshead. One important factor was climatic conditions. The ideal climatic condition for draining molasses was cool, dry weather; the worst weather for this process were periods of warm, humid days and nights. During the latter, it was perfectly possible for sugar and molasses to drain from the cask simultaneously resulting in little more than a worthless sticky mass on the floor of a warehouse or in the hole of a ship.

When packed with sugar and molasses a hogshead usually weighed about 1100 pounds. After the molasses had drained, that weight was reduced to about 800 pounds (or less, depending on the molasses content of the sugar). Quite naturally, such weight would have been difficult to move and transport overland, especially in view of the condition of what passed for roads in antebellum Louisiana. Thus, a plantation's sugar warehouse was usually quite near a navigable stream. Thus, when time came to ship the hogsheads of sugar to market, they were usually rolled from the warehouse to a waiting flatboat, schooner, or steamboat, loaded aboard (where the molasses-draining process might continue), and dispatched to markets in New Orleans and on the East Coast.[37]

The bulk of Louisiana's antebellum sugar was transported by water (as is much of the present-day crop) because it was the cheapest means of transportation. During the antebellum years, Louisiana had few railroads and even fewer regular roads. Since most plantations faced waterways, many planters had wharfs built to accommodate waterborne traffic, especially steamboats. After the War of 1812 and down to the outbreak of the Civil War, steamboats plied Louisiana's waterways, on either a regular or irregular schedule, stopping at plantation wharfs to deposit passengers and goods and/or take on passengers and produce. Depending on circumstances, a steamboat could arrive at a plantation wharf during the middle of the night, alerting everyone of its arrival by blowing its whistle. The entire plantation community would get up and go down to the wharf to begin loading and unloading the boat, a process that might require many hours.

A planter might sell his sugar directly to the steamboat captain or to a sugar speculator aboard the steamboat. More often than not, however, the planter paid the steamboat captain to deliver his sugar to the planter's factor, usually in New Orleans, but also possibly on the East Coast. A factor (or broker) was the planter's agent in a market city who was authorized by the planter to purchase or sell goods in the planter's name. Some planters, however, owned coastal schooners and took on the responsibility of shipping their sugar production directly to merchants in the market cities, particularly New York City and Philadelphia. The price of sugar on the East Coast was, on average, higher than that in the New Orleans sugar market, but if prices on the East Coast were lower than those in New Orleans, the result was that the planter with East Coast marketing ties lost money. Even higher prices on the East Coast did not necessarily mean a profit for the planter. Shipping and storage costs could absorb the difference between the higher East Coast price and the lower New Orleans price.

Marketing, therefore, was as important to the sugar grower as was cultivation, milling, and shipping. Consequently, there developed between the planter and his factor a special relationship. After the planter's sugar was taken aboard a vessel for transport to New Orleans, it was consigned to the planter's factor. It was the factor's responsibility to sell the sugar at the right time and for the best price.[38] In some circumstances, the factor may have had the sugar sold by the time it arrived in New Orleans, and therefore it needed only to be delivered to the buyer. Usually, however, the sugar was sold at auction along with the production of other sugar plantations. Prospective buyers would examine the sugar for sale, and, based upon grade, would offer purchase prices. It was in the best interest of the factor to secure the highest price for his client's sugar because he was paid a commission, normally 2.5 percent of purchase price.[39]

The buying and selling of sugar in the New Orleans market usually occurred in March, a time of the year referred to as "the season." People were said to be "spending the season in New Orleans," a time which usually coincided with pre-Lenten festivities. Planters and their families would go to New Orleans for the season to handle business affairs, make purchases for their plantations, examine the latest fashions, visit with other planters and their families, attend the horse races, dine, dance, and generally have a good time. In a few short weeks the season ended, and planter and family returned home to begin again the sugarcane cycle which would hopefully bring them back to New Orleans in a year.

Once the factor had disposed of his client's sugar, hopefully at a reasonable profit, he assumed his role as his client's purchasing agent. Usually on instructions from the planter during his New Orleans sojourn, the factor bought such things as tools, barrel staves to make hogsheads, sugar boiling and milling equipment, and a host of other items for the plantation. He would purchase domestic items for the plantations such as furniture, books, cloth, china, imported products, especially tea and coffee. To compensate the factor for these services, the planter paid him an amount based upon a percentage of the total of purchase prices.[40]

Factors often performed other services for planters and their families. When the planter visited New Orleans, he usually lodged at his factor's home, or the factor secured lodging for him. In some cases, the factor took on the responsibility of looking after the planter's children who were sent to the city for educational purposes. Many factors served as bankers for their clients. The nature of the sugar industry required the planter to put out large amounts of capital in advance of producing the annual crop. Whenever there was a capital shortfall, the planter would make the necessary loan from his factor on the assumption that the forthcoming crop would produce a profit from which the loan and interest could be repaid. Because sugar production was unpredictable, sufficient profits might not be forthcoming for the planter to handle all of his obligations, and thus the factor frequently "carried" the planter for many years in the hope that an exceptional sugar crop would come along with accompanying large profits. A planter, therefore, had to keep a sharp eye on expenditures because so often he was living on the brink of financial disaster.[41]

Factors were not the only source of credit for planters. There were two more, banks and other planters. There were established during the antebellum period several planters' banks. Planters bought stock in the bank, and then when a planter needed a loan he was fairly certain of getting it from his banking institution. Normally the planter could secure a loan with an interest rate of 6 percent per annum. This arrangement was good so long as the bank was called upon to make a limited number of loans in any given period of time. If, however, disaster struck the cane-growing region, such as hurricane or an unusually cold harvest season that killed the cane in the fields, then the banks would be besieged for loans, not all of which requests could be honored. In such circumstances, planters could only turn to other planters elsewhere in the hope of securing a loan to keep their operations going.[42]

A tactic of some planters to avoid financial disaster on a plantation was to own several plantations, not necessarily all sugar plantations. The Wilkins family, for example, large sugar producers in the Teche country, also owned and operated a tobacco plantation in Kentucky and a cotton plantation in

Mississippi. Thus, if a hurricane or cold destroyed the sugar crop, the Wilkinses could fall back on income from their other plantations to restart their sugar operation. Another tactic used by planters to avoid disaster was to divide their fields between sugarcane and another commercial crop, such as rice or cotton. Again, if sugarcane failed, the planter could expect to survive the crop year from the proceeds of his rice or cotton crop.[43]

If uncontrollable variables such as bad weather, crop disease, pests, mechanical failures, labor shortages, and creditors were not enough to test the resolve of a sugar planter, he also knew he had to face stiff competition from other sugar-producing areas of the world. Louisiana did not produce enough sugar for the consumption demands of the country. Even with the use of slave labor in Louisiana, it was still less expensive to produce sugar in the British, French, and Spanish colonies of the Caribbean Sea and Latin America. Cheap sugar was constantly being imported into the United States.

Occasionally during the antebellum years, the federal government imposed tariffs, or customs charges, on imported sugar. Between 1833 and 1846 the tariff was 2.5 cents per pound, with the result of equalizing the prices of imported and domestically produced sugar. After 1846, the tariff was changed to 30 percent ad valorem of a pound of sugar. This change together with a reduction in the price of imported sugar could have spelled doom for the Louisiana sugar industry because the price of imported sugar was then just slightly above the domestic price. The change in tariff resulted in an uproar from planters. Over the next fourteen years, the domestic price of raw sugar averaged 6.25 cents per pound, a price sufficient enough to let planters make a small profit.[44] As a result, the Louisiana industry survived and continued to grow slowly.

Throughout the two-hundred-year history of the Louisiana sugar industry, there has been a constant factor working not only for the survival of the industry but also for its expansion, a factor over which neither politicians nor foreign competitors have any control: advancing technology. Technological progress was first apparent in the milling operation. As noted above, change first came in the construction of rollers to crush cane in order to retrieve more juice.[45] An enterprising planter before the Civil War developed a mill with a double set of rollers, but his innovation would not be widely adopted until the era after the war.[46]

A major problem with rollers until the application of steam power and even afterward was that frequently cane was fed into the rollers too fast and the rollers would become stuck or "choked." If the mill did not stop soon after the rollers stuck, the pins on which the rollers revolved might break, thus causing the mill to stop operations until repairs could be made. Repairs could be costly in terms of time and money. To address this problem, improved rollers began to appear in the 1820s, and by the time steam power was being applied to the rollers they were much heavier, constructed of metal, and capable of extracting more juice from the crushed cane. More juice meant, in the long run, more sugar.[47]

Steam power was first introduced to the Louisiana sugarcane industry in 1822. By 1827 there were eighty-two steam-powered mills and 226 animal-powered factories. By 1843 there were 408 steam-powered mills, but the number of animal-powered mills had also increased, to 354. What these figures reflect is the fact that the sugarcane industry was expanding rapidly during these two decades. In 1831 a steam-powered mill cost about $4500, but this price did not inhibit planters from investing in the new technology. By the beginning of the Civil War, 1,027 of Louisiana's 1,291 sugar mills were steam powered.[48]

The adaptation of steam power contributed to the rapid depletion in many areas of forests as wood was the main source of fuel for firing the boilers. The open-kettle method of sugar production had consumed large amounts of firewood, usually about two to three-and-one-half cords of wood per hogshead of sugar produced. The use of boilers to create steam for the mill engines consumed even greater amounts of wood. The declining firewood resources coupled with increasing costs for supplies caused many planters to begin importing wood and coal from outside the sugar region. This, however, soon proved to be a costly solution to the problem of firewood shortage and led some planters to attempt use of dried bagasse as fuel. They soon discovered that the major drawback to the use of bagasse was that it had to be dried and stored, a costly process in itself. Then, in 1854 Mose Thompson developed an improved furnace which could burn wet bagasse as fuel. Within a few years many plantations were installing the new furnace, but it was not widely used in the industry until after the Civil War.[49]

One of the most widely used evaporator designs was the one developed by Norbert Rillieux, a free man of color from New Orleans. He was educated in France and was particularly adept in physics and mathematics. Between 1830 and 1832, while in Paris, Rillieux developed his evaporator. The design for which he received a patent in 1843 was a multiple-effect evaporator. The apparatus was made up of two bodies (or multiple effect evaporators) in which the sugarcane juice was boiled. Steam that was generated from the first body was used to heat the second body, which helped to reduce fuel consumption. In 1843 Rillieux set up his invention on a Louisiana sugar plantation, but it was not particularly successful. After reworking his design and adding to it, Rillieux received another patent in 1846. This time the apparatus contained three bodies. This design was more successful and by 1864 many sugar mills were using it. The Rillieux evaporator produced a superior quality grade of syrup, a major advance in Louisiana sugar production.[50]

The most exciting development in the milling aspect of the sugar industry prior to the Civil War was the invention of the vacuum pan method of producing sugar. The earliest vacuum-pan design used in Louisiana was the one developed by Thomas Morgan on the plantation of Valcour Aime in 1832. A different design which used two vacuum pans was developed in the 1840s by the French chemist E. Degrand. This vacuum pan design was adopted by some Louisiana planters in the 1840s. The use of these early vacuum pans was only spotty in the Louisiana sugar region. In 1843 only five mills were using vacuum pans. Ten years later 53 of 1,481 Louisiana sugar mills had adopted the new inventions.[51]

The purification process of sugarmaking was also advanced during the antebellum years. In the early years of the industry, lime was added during the boiling to clarify the sugar juice. As time passed, sugarmakers learned to filter the juice through boneblack and other minerals to clarify it. Further experiments resulted in findings that other minerals when mixed with lime purified sugarcane juice better than lime alone. By the 1840s experiments were taking place using bisulfate of lime to clarify, and in the 1850s it was discovered that sulphur fumes clarified sugar juice better than any other method used to that time.[52] Finally, just before the Civil War a few sugarmakers began using the French and German designed centrifugal methods of separating sugar crystals from molasses.[53]

In addition to these developments and research and experimentation there were several people in Louisiana in the two decades before the Civil War who were calling for agricultural reform. The foremost of these agricultural reformers was James D. B. De Bow, editor and publisher of *DeBow's Review*, an influential agricultural journal in the Deep South. Articles in the *Review* dealt with all facets of Southern agriculture, but articles on sugarcane production predominated in the journal.[54] DeBow published articles on many subjects, including rotation of crops, diversification of crops, insect and weed control.[55] The fact remains, however, that most of the technological advances of the antebellum era occurred in the milling process. Advances in cultivation would have to wait until later in the nineteenth century.[56]

III

Politics and the Civil War: The End of an Era

The use of African-American slaves as laborers in the sugar-growing region of Louisiana was widespread. Since sugar production required a large number of laborers, slaves were engaged in almost every aspect of the industry. Most sugarcane was grown on a plantation, a term which was liberally defined as any farm with 500 or more improved acres or one that had fifty or more slaves laboring on it. Using this definition of a plantation, only 40 percent of Louisiana's sugar growers could be classified as planters or plantation owners in the year 1850. By 1860 over 70 percent of sugar growers could be classified as planters, the increase owing to the prosperity generally found in the sugar industry during the 1850s and a consolidation of small production units into a larger unit. Even though plantations were continuing to increase in size and number of workers, the number of small farmers continued to be numerically large. Moreover, not all slaveholders were large slaveholders. Small farm owners also owned slaves, but half of all small slaveholders owned less than ten slaves. The small farmer, as a class, owned about 16 percent of all slaves in Louisiana. It was fairly easy to calculate the number of slaves needed for the amount of land under cultivation. It was generally recognized that a slave, in the prime of life, could till, cultivate, and harvest approximately five acres of sugarcane.[57]

Owning slaves was a huge financial investment for plantation owners. In the late 1820s, a field hand in the prime of life cost about $600. By the 1850s this price had escalated to between $1,200 and $1,500. A slave with skills, such as blacksmithing or carpentry, by the 1850s could cost between $2,000 and $3,000. Because of the increasing price of slaves, many slaveholders rented slaves to planters as the need arose. Some planters even experimented with free white labor, but these attempts failed because planters found that they could not control the work habits of white laborers to the extent that they controlled the lives of black slave laborers.[58]

The routine of slaves on sugar plantations was well regulated and changed little from year to year. Slave owners treated their slaves well enough to keep them working at a desired level. Slaves were housed, fed, and clothed in a humane fashion, for a slaveholder knew that a sick or hungry slave could not last long in Louisiana's sweltering summer heat or in the damp cold of a Gulf Coast winter. Sick slaves were often treated by the plantation owner's family or by local doctors. The Wilkins family was known to have a doctor come in once a year to give each slave a physical.[59]

The realm of politics was the exclusive preserve of white males and was usually dominated by the planter class. During the first quarter of the nineteenth century, Louisiana politics was closely related to the national Democratic Republican party. Later, under the leadership of Andrew Jackson, the party became known simply as the Democratic party. Opposition parties came an went, but one of these anti-Jackson parties, the Whig party, would exert large influence in Louisiana. The Whig party, organized in 1832, was composed of conservative followers of Jackson who had become disenchanted with the Democratic leader, and anti-Jackson people who had never bought into the president's politics. The Whig party's platform was a combination of then current political ideologies, but, for the most part, the party was in the political center.

The Louisiana Whig party did not have much to say about slavery except that it was necessary and should remain. The party was, however, a strong advocate of protective tariffs for agriculture, which was particularly appealing to sugar growers. The Whigs were also ardent supporters of internal improvements, such as roads and bridges and railroads, again an attraction for the sugar industry. Planters were also drawn to the party by its support for the formation of a national bank which would provide means of obtaining easy credit.

The Whig party thus emerged as the conservative party of Louisiana, the party of the sugar planters. The Whigs would control state politics into the early 1850s. The party, however, was to be a factor in national and state politics for only a few years, going into decline as a result of the breakdown of the political compromises which had given it birth. Whig supporters in North and South began to part company over the slavery issue.[60]

Slavery and the planter society that developed around it was the Southern contribution to the development of sectionalism in the United States. Northern industrialists and the region's commercial interests, determined to dominate the national economy, thus made their contribution to sectionalism. Louisianians were not immune to the growing sectionalism. Many of the state's residents were convinced that their political and economic interests were being threatened by the northern industrialists and their merchant allies. Fear of their situation caused many Louisiana planters of the 1850s to refer to themselves not as Americans but as Southerners. On the other hand, many of the state's white farmers and planters and people in other occupations were completely opposed to slavery and wanted to see the Union preserved, but people of this bent were in the minority. Finally, during the 1850s, the slavery debate in Louisiana and across the nation was transformed from a moral and religious issue to a political one.[61]

The presidential election of 1860 proved to be the flash point in the growing political crisis of sectionalism. Four candidates, reflecting various positions on the slavery issue, vied for the office of president. Abraham Lincoln, candidate of the Republican party, was elected. Many Southerners believed that if Lincoln were elected, he would abolish slavery and destroy the Southern "way of life." There were suggestions on all sides that if Lincoln won the only course for the South was secession. With Lincoln's election, South Carolina left the Union in December 1860--Louisiana followed in January, 1861.[62]

On January 7, 1861, an election was held to select delegates from across the state to decide the matter of secession. A majority of the delegates elected were in favor of immediately leaving the Union. A few compromise delegates were elected, but they could not halt the rising secessionist tide that was about to engulf Louisiana. On January 23, the convention voted 113 to 17 to secede. Three days later Louisiana left the Union.[63]

Events moved rapidly. Richard Taylor, son of the late president Zachary Taylor and a future Confederate commander, urged the creation of a military defense plan for the state. Volunteers were needed, and many men answered the call to arms, including planters and their sons. Anyone looking for war did not have long to wait, Fort Sumter, a federal stronghold, fell to besieging Southern forces on April 13, 1861. During the summer of 1861, sugarcane planters interrupted their preparations for the fall harvest only to hear and talk about the war in the East. Prospects for the harvest appeared good. It was better than that, 1861 was a great year for sugar production. The Louisiana crop resulted in 459,410 hogsheads—the largest to date.[64]

In early 1861 Louisiana joined the Confederate States of America, and sugar planters soon realized that they would face some of the same political problems they had faced under the laws of the United States. Sugar planters still favored agricultural tariffs, while other agricultural industries were opposed to them. Another problem, more related to the war, was the reduction of the white male presence in the sugar region as more and more left the state for the battlefields of Virginia. The absence of so many white males gave rise to fears of slave uprising, causing the number of slave patrols to increase. But a far greater problem was in the making as a result of so many troops leaving the state, the Louisiana Department commander, General Mansfield Lovell, found the state to be in a perilous defensive position.[65]

Adding to Lovell's concern was the fact that in the late months of 1861 the Federal navy began a blockade of the coast of the Confederate states. Although most sugar planters believed that no financial harm would come from the economic blockade, they were soon to learn differently. Sugar prices quickly dropped as planters found it impossible to ship the 1861-62 sugar crop. But the blockade would be a minor problem compared with the forthcoming invasion of the sugar region by the Federal army.[66]

In March 1862 a Federal fleet under the command of Admiral David Farragut took up a position off the coast of Louisiana. Obviously, the Federal plan was the capture of New Orleans, then the South's largest city, and subsequent control of the Mississippi River, thus isolating the western Confederate states from operations in the East. With only a rudimentary military defense, New Orleans soon fell to the Federal assault, led by Farragut and Major General Benjamin Butler. On May 1, Butler and his troops occupied the city.

The presence of Federal troops in New Orleans eliminated Confederate control of most of Louisiana's sugar region for the remainder of the war. The threat of Federal occupation of the region also caused many planters and their families to flee to Texas. From their base of operations in New Orleans, the Federal army began its invasion of the sugar country. In the fall of 1862, General Godfrey Weitzel led the invasion of the Lafourche country. On October 24 Weitzel landed troops near Donaldsonville and moved south along both banks of Bayou Lafourche. Some plantations were found to be in good condition, others had been abandoned and looted.[67] The social fabric was beginning to unravel.

General Butler, in an attempt to normalize life in the occupied areas, allowed all planters who remained loyal to the Union as well as former Confederate sympathizers who would take an oath of loyalty to resume sugarcane cultivation. The planters who resumed production faced many hardships. Wartime conditions caused the credit system of the sugar region to fall apart, ruining many planters, factors, and banks in the process. Another problem for planters, one that would plague them for many years to come, was the lack of a reliable labor force. When Federal forces invaded the sugar region, thousands of slaves left the plantations and crossed into Federal occupied territory. During the fall of 1862, there were about 10,000 black refugees in New Orleans. As the problem of refugees grew greater, General Butler decided that many of the blacks would be returned to their plantations, others would be employed by Federal forces, and still others would be fed and housed in New Orleans. General Butler developed a system in which planters would get the labor they needed and the blacks would receive food and shelter. The laborers would work ten hours per day for twenty-six days a month, and the planters would provide housing, food, clothing, and medical care. This system was to remain in place even after the war. Although this arrangement worked fairly well, planters of the sugar region did not react enthusiastically to it.

In December 1862, General Nathaniel Banks replaced Butler as commander of occupied Louisiana. Sugar planters complained about the wage system to the new commander; however, no one offered a substitute plan outside of the resumption of slavery. Banks changed the wage system slightly, now requiring planters to make annual contracts. Laborers would be paid a monthly wage, or they would receive one-twentieth of the crop at the end of the season. Planters continued to complain that laborers did not work as hard under the wage system. The labor situation was further complicated when, in January 1863, President Lincoln issued the Emancipation Proclamation, freeing all slaves in Federal occupied areas of the South. As the status of laborers remained confused, the economy of the area continued to deteriorate.[68]

The labor problem was only one of many facing the wartime cultivation of sugarcane. A serious problem developed as the number of draft animals in the sugar region declined. Mules and horses were stolen by fleeing slaves and by poor whites. Confederate and Federal military forces frequently confiscated the animals. The problem was never adequately addressed during the war years. As a result planters sometimes began to share their draft animals. Still another problem for planters was securing experienced overseers and sugarmakers. Many planters knew little or nothing about sugarcane cultivation or the making of sugar. Another problem was the destruction of plantation buildings, including sugarhouses, by the warring armies. Finally, wartime inflation caused the prices of everyday necessities to go beyond the ability of most people to pay. Planters and others associated with the sugar industry did what they could to overcome these hardships, but many plantations and sugar-related businesses did not survive the war intact.[69]

By 1863, Federal forces were in control of Louisiana from Port Hudson on the Mississippi River southward to the Gulf of Mexico. In the spring of that year General Ulysses S. Grant invaded Northeast Louisiana in preparation for the siege of the Confederate stronghold at Vicksburg. By July the two Confederate bastions on the Mississippi, Vicksburg and Port Hudson, fell to the Federal army. Later that year, the Federal army again marched through the Teche country on their way to attack Confederate forces in the northwestern part of the state. Thus, for all practical purposes, the sugar region of Louisiana was completely in Federal hands.[70]

Even though there were widespread military maneuvers in the sugar region during 1862 and 1863, a sugar crop for the 1863-64 season was planted and harvested. The crop resulted in a disappointing 77,000 hogsheads. Since most of the past year's crop had been grown from ratoons, seed cane was in short supply in 1864. To secure a supply of seed cane for the upcoming planting season, General Banks ordered all planters to save one fourth of their harvest to be used as seed cane. Some planters sold their entire harvest for seed cane instead of milling it, hoping that the seed would bring a better price than the raw sugar.[71]

Military operations in Louisiana in 1864 were mainly in northern Louisiana. Alexandria fell to Federal forces in March. General Richard Taylor's Confederate forces began retreating toward Shreveport. With the Federal army in pursuit, the Confederates turned and made a stand at Mansfield on April 8. That battle was inconclusive, even though Federal forces retreated to Pleasant Hill. The Confederates again attacked, and Federal forces again broke lines and began retreating toward Alexandria. Taylor eventually pursued the Federal forces, but his force had been so reduced in size that it could only harass the enemy. The Federal's Red River Campaign had come to an end without achieving the goal of driving the Confederates out of Louisiana. At the same time, however, the campaign had succeeded in destroying or severely damaging most of the sugar plantations in the Red River country.[72]

During the remainder of 1864, there were only a few minor skirmishes between Federal and Confederate forces. The Confederate-controlled area of the state held a gubernatorial election that year, and Henry Watkins Allen was elected governor. Confederate state government, however, remained almost totally ineffective because so much of the state was occupied by the Federal army. In the occupied area of the state Michael Hahn was elected governor. Under his direction the Constitution of 1864 was drawn up. The new constitution outlawed slavery in the state, sugar planters were now forced by law to use the contract wage system.[73]

By the spring of 1865, the war in Louisiana had all but ended. The sugar crop for 1864-65 produced only 18,000 hogsheads of sugar. There were only 188 sugarhouses operating that season. Sugar planters blamed the poor harvest on the lack of reliable labor. They knew the situation would only worsen if the war continued for much longer, but it soon ended. In April, General Robert E. Lee surrendered his Confederate force to General Ulysses S. Grant. Other Confederate generals followed Lee's example and surrendered. General Kirby Smith surrendered Confederate forces in Louisiana on June 2, 1865. By this time local government in the Confederate-held areas of the state was almost nonexistent. Governor Henry Watkins Allen fled Louisiana to a self-imposed exile in Mexico.

The Civil War had come to an end, but the cost of the four-year contest had been staggering, especially to the sugar region of the state. Total investment in the sugar industry in 1860 had been about $200,000,000, but by 1865 that amount had fallen to $70,000,000. In terms of dollars, more property had been destroyed in Louisiana than in any other state. Despite these losses, despite the ravaged plantations, and despite the fact that the traditional labor force was scattered and confused, sugar planters, especially those involved in the day-to-day operations of the industry, were optimistic that there was still a future for sugar in Louisiana.[74]

GLIMPSES
OF THE
LOUISIANA
SUGAR INDUSTRY
OF THE PAST

A PHOTO ESSAY

Etienne Boré (above) Norbert Rillieux (below)
Louisiana Sugar Pioneers
(Both images courtesy of The Historic New Orleans Collection:
Boré, Acc. No. 1974.25.27.46; Rillieux, Acc. No. 1974.25.13.198)

Planting Sugarcane Long Ago

Sugarcane Sprouting in the Warm Sun of a Louisiana Springtime

Cultivating Young Sugarcane (above) Cutting Cane (below)

A Cane Cutter Takes a Break

Loading an Old-Fashioned Cane Wagon

Loading Cane with an Early Mechanical Loader (above)
The Long Haul to the Mill (below)

An Early Experiment in Mechanical Sugarcane Cutting

Mechanization Comes to the Sugarcane Field

Loading Cane for Trucking to the Mill

Trucks, Wagons, and Railroad Cars Bring the Harvest to the Mill
Waiting to Unload (below)

A Typical Nineteenth-century Sugar Mill

A Kettle Used to Boil Sugarcane Juice

MARIA CENTRAL FACTORY.

Billeaud's Sugar Factory in the 1930s

IV

Rebirth: The Beginnings of Modernization

The era between the Civil War's end and the first decade of the twentieth century were years during which the sugar industry revived and there occurred changes leading to its modernization. The era from 1865 to 1910 might be divided into two major time periods based upon political developments within the state. The first of these was the period of Reconstruction, which lasted approximately fifteen years and which can be subdivided into two distinct phases: the Presidential phase and the Congressional or Radical phase. The second span of years within this era might be characterized as the period of Bourbon rule, running from 1876 down into the first decade of the twentieth century.

Reconstruction in Louisiana began almost as soon as the first Federal troops invaded the state in early 1862. It had become President Lincoln's policy to restore the rebellious states to full membership in the Union as soon as possible. For this to occur, however, the states to be readmitted had to create a new state government loyal to the Union. Whenever 10 percent of the males in a rebel state who had voted in the election of 1860 took an oath of allegiance to the Union, the process of drawing up a new state constitution and forming a new government could begin. This, basically, was the plan of Presidential Reconstruction.

The Federally occupied areas of Louisiana had a new government as early as 1864 which agreed to draft a new state constitution in which slavery was abolished and the state agreed to pay all debts incurred by the Confederate regime. After the cessation of hostilities, a new state election for governor and legislature was held according to the provisions of the 1864 constitution. The election returned to the legislature a majority of ex-Confederates who refused to guarantee the rights of African-Americans and also refused to ratify the Fourteenth Amendment to the Constitution, granting citizenship to ex-slaves. Even though the state had not lived up to its side of the bargain in Presidential Reconstruction, the moderate policies of President Lincoln continued after his death and well into the term of his successor, Andrew Johnson.[75]

By 1867, however, the Republican-controlled Congress was more and more inclined to listen to radical members who railed against the Reconstruction governments in the South, saying that they were once again controlled by the same elements who had supported slavery and secession. The immediate result was that Congress refused to seat many congressmen from the former Confederate states, basing their action on these states' refusal to ratify the Fourteenth Amendment.

Then, to force its will on these Southern states, Congress, led by the Radicals, passed a series of military reconstruction acts. The ten states which had not ratified the Fourteenth Amendment to the Constitution were divided into five military districts and occupied by Federal troops These troops would serve the dual purposes of insuring the recently freed blacks security in their voting and civil rights and propping up the Republican-controlled state governments that came to power through contrived elections.

The first of these Reconstruction acts was passed in 1867, and by its terms Louisiana and Texas were designated the Fifth District under the command of General Philip Sheridan. The military occupiers would register loyal white and black voters who would elect a new state government. After Louisiana passed a new state constitution and the state ratified the Fourteenth Amendment, Congress agreed to seat the state's congressional delegation and designate the state as an equal partner in the Union.

Throughout 1868, the Louisiana legislature complied with the demands of Congress. A new state constitution was drafted, containing a civil rights provision and mandating the creation of an integrated public school system. A new governor was elected that year—Henry Clay Warmoth, a carpetbagger and sugar planter. Four years later William Kellogg, a Republican, was involved in a contested election result, but became governor once President Grant recognized him as the duly elected governor.[76] Needless to say, many Louisianians chafed under the Republican regime and awaited the opportunity to cast the party from office.

That opportunity was not long in coming. In 1876 the voters of Louisiana elected a new governor and participated in the election of a new president. The gubernatorial race was between Democrat, and ex-Confederate, Francis T. Nicholls and Republican Stephen B. Packard. The election, involving widespread voter coercion and election fraud, resulted in both candidates claiming victory. With no resolution of the contest, both men set up a government in the state. The only question was: how would the impasse be broken?

The answer came in the presidential election of 1876. Democrat Samuel J. Tilden was pitted against Republican Rutherford B. Hayes. Tilden won the presidency in the popular vote and in the electoral college, but Republican leaders were determined to hold on to power. On the morrow of the election the Republicans claimed that the presidential election had been filled with fraud, particularly in the former Confederate states still under federal occupation—now only Louisiana, Florida, and South Carolina. Thus the Republicans challenged the presidential election returns in those three states and in the state of Oregon. By claiming the electoral votes of all four states, the Republicans would have 185 electoral votes for Hayes to the Democrats 184 votes for Tilden.

But who would decide whether the four states had truly gone Republican in the popular vote. An Electoral Commission was established to decide the issue. The Commission was composed of five House of Representatives Democrats, five U. S. Senate Republicans, and five supposedly impartial members of the U. S. Supreme Court, even though everyone knew that three of the five Court representatives had Republican sympathies. As everyone knew would happen, the Commission declared Hayes the winner of the presidential contest, but not before Democrats, as a price for remaining quiet on the issue, made Hayes agree to withdraw federal occupation troops from the Southern states.

Thus, as a result of the 1876 presidential election, Louisiana's white male population, a majority of which was Democratic in sympathy, was left in control of the state. With the removal of federal troops and the installation of Nicholls as governor, Reconstruction in Louisiana had ended, and the so-called Bourbon era was dawning.[77]

The Bourbon era can be described simply as a return to the values and many of the lifestyles of the antebellum years, with one difference—a determination to push for industrialization, not only in Louisiana but across the South. With the coming of Bourbon rule, most of the political and civil rights of African-Ameicans were reduced or obliterated. Blacks, though still free from the bonds of slavery, were at best members of a vast peonage stretching across the South. The Bourbons held political and economic sway until the end of the nineteenth and beginning of the twentieth centuries when men favoring progressive reforms began winning gubernatorial elections. In Louisiana, the Bourbon era is said to have ended with the election of Huey Long as governor in 1928. The fact is, however, that Long's greatest accomplishment was replacing the Bourbon oligarchy with one of his own.[78]

Throughout the years of Congressional Reconstruction and into the early Bourbon era, the routine of the sugar region of Louisiana remained much the same. Planters were busy rebuilding their plantations and trying to become accustomed to the new free black labor system. In contrast with other agricultural pursuits in the South, the sugarcane industry continued to adhere to the old plantation system of the antebellum years, albeit without the institution of slavery. As many Louisiana planters lacked capital required to run a plantation and sugarhouse, Northerners began investing in sugarcane cultivation. They bought plantations and sugarhouses or simply invested in ones that were already in production. In either case, they, too, soon came to agree with their Louisiana counterparts that the supply of black labor was not reliable.

The problem brought to the sugar industry by the new labor system forced planters to begin looking for new sources of labor. The problem arose from two sources: the inability of the planters to overcome a desire to control their laborers as they saw fit, and the desire of the blacks to exploit their new-found freedom. Before mechanization, as noted, sugarcane production was labor intensive, particularly large numbers of workers were needed for cultivation and harvesting. This large force had to be available when needed; reliability was a key factor. But many among the native black population misunderstood their new freedom to mean that they were free from doing manual labor for others. Many black laborers would enter into contracts with planters to work through the sugarcane season only to break the contracts for any one of a host of reasons, including taking a job with a planter who was paying slightly more for the daily or monthly wage. Even though many black laborers entered and fulfilled the terms of the contract, so many did not that planters began looking for alternate supplies of labor.[79]

One of the earliest attempts of sugar planters to gain a stable work force involved Chinese laborers. Louisiana planters brought in Chinese workers from California, the Philippine Islands, and China. The planters were conscious of the good work ethic of the Chinese and believed that these workers would help reinvigorate the sugar industry. Soon, however, the planters were to be disappointed. The Chinese laborers did work well in the cane fields, but all too soon they left the plantations in search of less demanding work and higher wages. The problem of a reliable labor force was not solved.

Some planters began directing their recruiting attempts to immigrant labor from Europe. So many European immigrants were coming to the United States in the latter half of the nineteenth century that it was felt many of these would take jobs in the sugar fields of Louisiana. Planters dispatched recruiters to Northern cities where the immigrants had congregated. At first the recruiters were successful, especially among Scandinavians who signed up for work in Louisiana. After a short time in the sugar fields, the Scandinavians, like the Chinese before them, began leaving for higher wages and better work. Many left simply because of the Gulf Coast climate.

Learning a lesson by trying to hire laborers from northern Europe, planters turned to another group of immigrants, a group which they felt might better endure Louisiana's summers—the

Italians. Initially, the Italian laborers worked well, but, in time, they failed to provide the reliable labor force that the owners desired. After a year or two in the cane fields most Italians left and joined the communities of Italians in New Orleans and other Gulf Coast cities.[80]

Many planters were frustrated by their inability to secure a reliable work force. There was little more to do than to try to hire black laborers and reduce production to fit the size of the labor force. Other planters, however, began to suggest that sugar interests should band together to address the chaotic conditions existing in the sugar region. These planters, when organized, would have a positive effect on the production of sugar in Louisiana well into the twentieth century.

The organization of planters which was founded on November 20, 1877, was called the Louisiana Sugar Planters' Association (LSPA). Its foremost purpose was the furtherance of the sugar industry in Louisiana. The Association would concert efforts to secure and maintain a reliable labor force; it would advocate the use of science and technology throughout the industry; and it would promote protection of domestically produced sugar through tariffs. Members of the Association came from all parts of the sugar region, people who had had long experience with sugar production and people who were only then trying to become involved in the industry. Many of the wealthy sugar growers of the state were prominent members of the Association, but wealth was not a qualification for membership. The only membership qualification was that a person have an interest in furthering the economic foundations of the industry in the state. Thus, planters of Northern or Southern nativity, veterans of Federal and Confederate military services, Republicans and Democrats joined forces to further the cause of the Louisiana sugar industry.[81] The LSPA is important to the history of the Louisiana sugar industry not because it was the first planter organization, which it was not, but because it would have lasting effects on the industry.

One of the LSPA's first projects was to aid planters in securing a reliable labor force for their plantations. The Association together with the state government issued pamphlets and flyers in an effort to induce immigrant laborers to come to

Louisiana. Owing to the fact that immigrant workers later proved not to be a reliable source of laborers, planters, with the help of the LSPA, combined and set maximum wages for field labor to insure that black laborers would not go to other plantations in search of higher wages once they contracted to a planter. These and other efforts tended to stabilize the availability of a labor force in the sugar region.

Another important contribution of the LSPA was the pursuit of technological and scientific advances for the sugar milling process. Planters realized that they needed to use progressive technology in an effort to reduce the cost of producing sugar. This would enable them to save money on the production side of the industry and also to secure higher prices on the marketing side through the marketing of higher grades of sugar. Planting and harvesting of sugarcane during the late nineteenth century remained much the same as it had been during the antebellum era. This aspect of the industry would not begin to change meaningfully until the early twentieth century and the coming of the tractor.

If field technology remained pre-industrial, the LSPA began urging planters to industrialize the sugar-making aspect of the industry. What technology that had come to the industry in the antebellum and immediate post-Civil War years was largely owing to founders who improved on machinery used in the sugar-making process. There were advances made in the manufacture of boilers, burners, kettles, vacuum pans, and centrifugals. Old-time sugarmakers and foundry men, who used rule of thumb skill in their trades were beginning to be replaced by college-educated engineers and chemists. These professionals quickly demonstrated to planters the value of new methods by producing higher grades of sugar using new techniques.[82] By the end of the nineteenth century many of the old ways of sugarmaking were disappearing in the face of the advancing industrial revolution.

The LSPA was also instrumental in inducing sugar people to use scientifically based data. Much of the early scientific data was gained from experiments made by scientists working in the European beet-sugar industry, especially in France and Germany. The Association now urged members to test varieties of sugarcane to determine which were best for their growing conditions. At the same time college-trained agrologists were testing soils to determine nutritional content and what type of fertilizers to apply for best results. Other scientists were experimenting with sugarcane varieties to determine those best suited to growing conditions in Louisiana.

The increased demand for trained chemists on Louisiana's sugar plantations prompted the LSPA to promote the idea of a state chemist. Once that position was created, the LSPA, with the help of the United States Department of Agriculture, lobbied for and succeeded in having set up a series of agricultural experiment stations in Louisiana for the benefit of all agricultural industries. The stations conducted field analyses and manufacturing experiments for sugarcane and other state agricultural industries and then published results.

A positive result of the experimental stations was the creation of the Audubon Sugar School in 1891. The school was opened on the grounds of the Audubon experimental station in a cooperative effort between the LSPA and the USDA, but was funded almost exclusively by the LSPA and other private funds. Students were offered a two-year curriculum which was developed in part by Harvey W. Wiley of the USDA. While enrolled, students would study the latest methods of sugarcane cultivation and manufacture. The Audubon school would become a major training school for the sugar industry, but, unfortunately, its success lasted only a few years. In 1896 the school was closed because of lack of funding. At almost the same time foreign experimental stations began massive research projects. The foreign activity in this regard was in reaction to the USDA's decision to cut back funding for agricultural research and experimentation.[83]

Although the Audubon Sugar School and the Louisiana experimental stations quickly lost their competitive edges, their contibutions to the sugar industry were many indeed. Had it not been for the LSPA, many of the advances made in the closing years of the nineteenth century would not have come about until several years later. The LSPA did not only focus its strength on the scientific and technological aspects of the sugar industry, but it also used its tremendous political influences to bring a better industry into the twentieth century.

Shortly after the organization of the LSPA, Louisiana sugar interests began to lobby for stronger tariff protection. This had been one of the original reasons for the creation of the LSPA. It was argued that high tariffs would keep wages high for black laborers, a convincing argument in almost any political circle. Moreover, many members of the LSPA wielded considerable influence in state government and exercised some influence in federal government circles. This political clout was not limited to a single political party, for the LSPA had powerful Democrat as well as Republican members. Such bipartisan political influence enabled the LSPA to lobby Congress for favorable tariffs for the sugar industry. The Association even achieved having one of its members appointed to the federal tariff regulation board, the group that recommends tariffs to the president and Congress. The LSPA was an effective political force well into the twentieth century.[84]

During the last quarter of the nineteenth century, with, and sometimes against, the insistence of the LSPA, the U. S. government enacted numerous tariffs and engaged in several treaties which affected Louisiana's sugar industry. In 1870, Louisiana sugar producers were protected by a 1.75 cents per pound tariff on imported sugar. This duty was raised to roughly 2 cents per pound between the years 1875 and 1883. But in June, 1875, a treaty was signed with the then independent kingdom of Hawaii, to ship sugar into the United States duty free. At the time Hawaii was producing only about 13,000 tons of sugar per year. By 1890, however, Hawaiian sugar producers were sending 140,000 tons per year duty free to the mainland. That year the McKinley Tariff somewhat changed the Hawaiian sugar producers' advantage. This tariff schedule placed raw sugar on the free list and put a small tariff on imported refined sugar.

The way the new tariff protected the mainland sugar industry was to include a sugar bounty for mainland sugar, both cane and beet. The United States government would pay 2 cents per pound for domestically produced sugar. The sugar bounty encouraged Louisiana's planters to increase their production, causing a small boom in the state's sugar industry. The McKinley Tariff was designed to operate for fifteen years; however, tariff matters soon took another direction. After Grover Cleveland's election as president in 1894, the McKinley Act was repealed and the Wilson-Gorman Tariff Act was passed. Among the provisions of the new tariff was the free listing of Hawaiian sugar, and all raw sugar imported would be charged 40 percent ad valorem. Refined sugar had a duty of .5 cent per pound. The Wilson-Gorman Act also repealed the domestic sugar bounty.[85] This latest tariff measure marked only the beginning of a series of problems for Louisiana sugar planters having to do with foreign competition and tariffs.

Internal events in Cuba, Louisiana's largest sugar competitor, should have commanded planter attention, but that attention was too narrowly focused on the repercussions of the McKinley Tariff repeal. Cuba, a Spanish colony since the early sixteenth century, became the scene of political unrest during the 1890s that led to an active movement for independence. Before that occurred, however, William McKinley was elected president, and in 1897 the Dingley Tariff gave some relief to Louisiana sugar interests. As the political situation deteriorated in Cuba, leading to the Spanish-American War, Louisiana planters began to fear that the defeat of Spain would result in the annexation of Cuba and the end of the Louisiana sugar industry.

The results of the Spanish-American War posed serious problems for Louisiana growers. In 1900, Puerto Rico, now a United States protectorate, received an 85 percent reduction in tariff rates, and in 1901 was granted duty-free status. In 1902, the Philippine Islands, which had been occupied by U.S. military forces, were granted a 25 percent reduction in duties. In 1909 the Payne Act allowed the Philippines to ship 300,000 tons of duty-free sugar to the United States, and in 1913 duty-free status was given to the entire Philippines sugar crop. In 1902 Cuba had received a 20 percent tariff reduction. Thus the new century brought with it few encouraging signs for the Louisiana sugar industry. Production in the state began to fall rapidly until it reached its lowest point in the late 1920s.[86]

Although there were many technological and scientific advances in the sugar industry between 1860 and 1875, the fact remains that the traditional way of growing sugarcane and processing it for sugar on the plantation went uninterrupted. Beginning with the last quarter of the nineteenth century, however, some planters began to understand that the practice of milling on the plantation kept production costs and efficiency down, and therefore growers began calling for a milling operation which would serve several plantations.

Here was born the idea of the central sugar factory. Like so many innovations, however, the central sugar factory had its supporters and opponents. Opponents pointed out that investment in such a large operation was risky for the would-be central factory owner. They also rightfully called attention once again to the poor transportation system of the state. Sugar growers would have to be convinced that the means of transportation to the central factory was available and practical.

Proponents of the central sugar factory believed that this approach to milling would provide greater efficiency and would increase profits for growers and millers. Advocates also pointed out that the cost of milling equipment was going beyond the means of many farmers. They believed that only corporations would be able to afford the latest milling and sugar-making equipment.

The central sugar factory was not long in coming. One of the earliest was the Vermilion Rose Hill Company which opened in 1879. This company purchased sugarcane from farmers and planters for use in its mill, and proved that transportation was not an overwhelming obstacle to the central factory idea. Another central factory, the Carencro Sugar Refinery Company, Ltd., claimed that in 1894 it purchased from local growers a total of 2,000 acres of sugarcane. By the mid 1890s the central factory concept of milling cane had become generally accepted in the Louisiana cane belt.[87]

Even so, some problems still remained. One was how would the mills pay growers for sugarcane purchased. Another problem was how to secure enough cane to keep the factory running at full, or near-full, capacity during the grinding season. The answers came when mills began contracting with growers to buy their entire production. Growers would be paid a certain price per ton of sugarcane based on the current price per pound of raw sugar on the New Orleans market. Other milling corporations, family or public, bought sufficient farmland to produce the cane necessary for their needs. One thing was certain, however, there must never be more cane available than the mills could grind in the October to early January season because cane without a home risked being left in the fields after a January freeze.

The effects of milling centralization were swiftly coming. In 1880, 1,144 largely plantation mills produced 136,745 tons of sugar. By 1889, a combination of 746 plantation mills (but far fewer in number) and central mills produced 143,745 tons of sugar. As the number of small plantation mills declined rapidly, the number of central factories with far greater sugarmaking capacity increased as did overall sugarcane production. In 1900, there were 300 sugar mills in Louisiana. By 1910 that number had dropped to 214, but these mills were producing 325,000 tons of raw sugar.

By the turn of the twentieth century, a major problem was developing for millers and growers. Field technology had not kept pace with mill technology. Mills were grinding so much more cane faster that farmers encountered increasing difficulty harvesting and transporting the supply required by the mill. Sugarcane was still being cut by hand, loaded onto carts by hand, and transported to the mill mainly through the use of draft animals.

Field technology was advancing but it would be several years into the future before growers began using tractors to plow and cultivate their fields and use tractor-drawn carts to get the cane out of the fields and to the mills. A major technological advance was on its way, the mechanical sugarcane loader. This was usually a combustion-engine device which would pick up the cut cane in large bundles and place these in a cart for transporting. Not only was the backbreaking work of loading carts obviated by the cane loader, but the device could load much more cane faster than could humans.

As the first three decades of the twentieth century passed, almost everyone involved in the sugarcane industry realized that production was now geared to how much cane could be hand cut in a season. That, of course, depended on the labor supply, and as long as the labor supply remained large few people were willing to invest money into planning and designing a practical mechanical cane cutter.[88] Nevertheless, experiments continued and by the 1940s the mechanical cutter had entered Louisiana's cane fields. A new era of production was born.

The major themes of the post-Civil War Louisiana sugar industry, down to 1900, were expansion, protection, and modernization. These three themes became focal points for the industry because it was understood that Louisiana sugar production had to become more profitable and more competitive. Outside forces, exerting huge influences on the local industry, were factors for modernization. Competition from other world sugar suppliers, beet and cane, spelled doom for the Louisiana industry unless all three post-war goals were attained.

The Louisiana industry was forced to change its ways or face extinction because of world competition. The resulting scientific and technological progress changed the face of the local industry. Milling and growing became two distinct operations, both increasing in tempo, but cultivation always in the wake of manufacturing.

On the political side, government tariffs did help to a degree, but at the same time the United States opened new doors to many competitors of the Louisiana industry. Without the many faceted assistance of such organizations as the LSPA and the USDA, Louisiana's sugar industry's move into the twentieth century would have been slow and may not have survived the crisis years of the late 1920s.

V

The Louisiana Sugar Industry in the Twentieth Century

At the dawn of the twentieth century, the Louisiana sugar industry was gearing up to the challenges which surrounded it. Many of the trends of the industry's first century would be repeated during its second century. Foremost among these were the periods of progress followed by periods of stagnation. Many challenges were ongoing such as adverse weather conditions, political and economic crises, war and depression, plant disease, world competition, labor shortages, and the need for further modernization.

Shortly after the turn of the century, the specter of foreign competition seemed to fade somewhat as European nations removed bounties on beet sugar. Unfortunately for Louisiana, however, a depression engulfed the industry from 1906 to 1914 as a result of falling prices, early freezes, floods, plant diseases, and labor shortages. Undaunted, the LSPA continued its fight for tariff protection, mechanization, and scientific advances.

Throughout the nineteenth century, the LSPA had used its considerable political influence to promote the Louisiana sugar industry. But, as noted above so many foreign sugar interests were tugging the arm of Congress that the LSPA's influence began a decline. The organization thus merged in 1922 with the American Cane Growers Association to form the American Sugar Cane League.[89] The new organization would have a positive impact on the state's sugar industry from that time down to the present.

As noted above, Congress passed a law in 1913 which allowed all or part of Philippine sugar production to enter the United States duty free. The next year, the year in which World War I began in Europe, Congress passed the Underwood-Simmons Tariff. This charged about a half cent per pound tariff on sugar imported from nations not on the duty-free list, and provided for the free listing of all sugar in 1916. Perhaps the outbreak of war caused the framers of the legislation to fear an upcoming sugar shortage, and thus the delay in implementing free listing. In any case, the free listing of sugar would mean the end of Louisiana sugar production. Louisiana growers feared the approach of 1916.

In that year, however, Congress, taking note of the chaotic situation in Europe, and going along with the suggestion of President Woodrow Wilson, voted to continue the 1.25 cents per pound duty on sugar. The domestic sugar industry, beet and cane, breathed a sigh of relief.

The average price of a pound of sugar in the fall of 1916 was 5.3 cents per pound. As the nation geared up for war, the sugar industry began to experience a labor shortage and with that shortage wages began to increase. The 1916 wages of cultivation and harvesting laborers was 80 cents and $1.25 respectively. By 1918, these wages had risen to $1.25 and $2.00 respectively. However, these wage increases did not last long because of changing postwar conditions in the industry, and wages began to drop off in early 1921.[90]

One reason why wages increased so rapidly was owing to government control of the industry after the United States entered the European war in 1917. All areas of the industry prospered during the period of wartime controls, but in ways that were not immediately apparent. Increased wages, for example, encouraged the industry to begin using more machinery in field operations. At war's end there were approximately 150 tractors in Louisiana's sugar fields, by the 1920 season, there were almost 500. The early tractors were not, however, a complete boon to the industry. The unusually hard work to be done in sugar fields caused many tractors to breakdown. Moreover, Louisiana's muddy field operations defied tractor tires to get through the muck pulling a loaded cart. Thus, the trend toward mechanization in the fields was slowed by persistent mechanical and technological problems. The extra time and cost of repairing tractors and getting them unbogged was thought by many growers to outweigh the advantages of mechanization.

When World War I ended in Europe in 1918, the nation and the world was facing a sugar shortage. The federal government therefore continued price controls on sugar into the middle of 1919. Then, with the end of price controls and with the fear of a worldwide sugar shortage, the price of sugar rose to 24 cents per pound. Such a price encouraged growers to plant more acreage in the fall of 1919 and to upgrade their operations with new equipment. Unfortunately for the growers, the high price of sugar did not last long. By August, 1920, the price of a pound of sugar had dropped to 12 cents; in October to 8 cents; and by December, with the realization that there was no worldwide sugar shortage, to 5 cents.[91] The consequence was, of course, massive financial collapse. Many growers in the 1920 to 1922 era filed for bankruptcy because they could not pay for acreage expansion and new equipment. Small farmers suffered to a far greater extent than their larger counterparts. But financial ruin was only the beginning of problems to be faced by the sugar industry in the 1920s.

Probably the economically darkest days of the Louisiana sugar industry occurred in the decade of the 1920s. At first, it was thought that there was a simple political solution to the industry's problems. As price collapse became obvious, Congress raised the duty on Cuban sugar to 1.6 cents per pound, and a little later, in 1922, raised it to almost 1.8 cents per pound. It was believed that these tariffs would help the industry get through temporary financial hardships. What was unknown was that Nature was now prepared to take a hand in creating new hardships.

During the 1920 growing season, the United States Department of Agriculture announced that the sugarcane disease called mosaic was found in sugarcane samples taken in Louisiana. Mosaic does not kill the sugarcane plant; rather, it attacks the leaves, weakens the plant, and thus makes it more susceptible to other plant-killing diseases. The disease, not easily detectable by any but the trained eye, had already reached epidemic proportions when it was discovered in Louisiana. Unsuspecting growers had accelerated the disease's spread by planting diseased seed. At the same time, funding cutbacks for experimental stations delayed the discovery of a cure for the disease or for developing mosaic-resistant varieties of sugarcane. Once funding was restored, the USDA and the state's experimental stations began to develop mosaic-resistant varieties of sugarcane, but not until the industry had been virtually devastated. By 1924 new mosaic-resistant varieties were being planted throughout the state's sugar belt.

Mosaic was not the only disease to affect sugar growers. During 1924 the disease anthrax killed almost 20,000 stock animals. Again, it was the small to medium-sized growers who were affected, for they still were using draft animals for cultivation of their sugar crops.

Added to the disease problems was the fact that climatic conditions did not favor high sugar yields. A drought in 1924, from the beginning of June to the end of August, was so severe that many growers pumped water from nearby streams and swamps in an effort to save their cane from dying. The next year, 1925, the harvest month of October was unusually wet. More than 12 inches of rain fell on the sugar region. The 1926 crop grown on 127,000 acres yielded only 50,000 tons of sugar. Mosaic was still taking its toll. In 1927 only 73,000 acres were planted in cane, and much of the cane region was inundated by the great Mississippi River flood of that year.

That year, however, marked the introduction of the POJ variety which was widely planted. By 1929 the worst was over, and that year's crop produced 190,000 tons of sugar. Louisiana's sugarcane industry had survived, and survival was owing mainly to the efforts of the USDA and the state's experimental stations. Through the introduction of new varieties of cane, and with the growers awareness that the old Cuban varieties were no longer profitable, the industry was on the road to recovery. As the 1930s dawned, growers were being urged by USDA and state experts to plant disease-resistant cane, use better fertilizers, and begin a crop-rotation plan.[92]

The trend toward centralization of mills that began in the late nineteenth century continued into the twentieth century. In 1912, there were 158 sugar mills in Louisiana, which produced 3.1 million pounds of sugar. As the number of mills declined, those remaining were increasing in capacity. In 1916 the Godchaux Reserve Factory was grinding 10,000 acres of sugarcane per season, which averaged out at about 1,800 tons per 24-hour day. But as the various problems of the 1920s set in, only 54 mills operated in 1926, and the problem of having a mill nearby to grind cane became a serious one for the grower. A sugarcane farmer could grow all the cane he wanted, but if there was no mill to grind it, or if a mill owner decided that there was not sufficient profit in operation and opted to keep the mill closed for a grinding season, the result was disastrous at best for the grower.

The response to this situation born of conditions in the 1920s as well as to the lack of mills in a sugarcane growing area of the state was the introduction of grower-owned milling cooperatives.[93] These mills, called co-ops, were owned by growers and processed all grower-owners' sugarcane production. If the capacity of the mill was great enough, the co-op would also agree to grind cane for non-members of the area. Buying in to the mill assured growers of a home for their cane. By the end of the 1930s there were 82 operating mills in Louisiana and six of these, mostly new mills, were grower cooperatives. As an example, the Iberia Sugar Cooperative, Inc., New Iberia, began operations for the 1937 season serving slightly more than 400 growers, 300 being members of the cooperative. In 1937 the mill ground 91,200 tons of cane. In 1938 it ground 144,900 tons, delivered to it from a radius of 15 miles.[94] As the century progressed, co-ops would play an ever-growing role in the manufacture of sugar.

The end of the 1920s, as noted, brought a ray of hope to Louisiana's sugar industry. But that ray was nearly extinguished with the coming of the Great Depression. The Depression occurred at a time when world sugar prices were low because of a world sugar surplus. The number of sugar mills continued to decline as the financial situation of growers deteriorated.

As paralysis gripped the industry the federal government sought to relieve the situation. In 1934 Congress decided to take control of the sugar industry. A quota system was introduced for the domestic production of sugar and for the importation of foreign sugar. It was this quota system which may have gone far toward saving the Louisiana sugar industry from extinction. Under the quota system, the entire sugar market of the United States was divided up between the various domestic producers and importers of foreign sugar. Each sugar producer would be given a specific percentage of the whole nation's sugar market. For example, in 1934, Louisiana's quota was 260,000 tons of sugar. This initial quota was to be divided among the sugar growers of the state who would each receive a specific quota to produce. Growers would agree to plant only that amount of sugarcane corresponding to their quota. In return, the federal government would pay the farmer a fixed sum per month to cover some of his costs. The federal government would raise the money by a processing tax on sugar. The system seemed to work well, but growers complained that their quotas were too small to warrant their investments of time and money. The issue of quotas, however, became moot when the Supreme Court declared the Agricultural Adjustment Act to be unconstitutional.[95]

Undaunted, President Franklin Roosevelt continued to advocate agricultural reform to fight the Depression, and, in 1937, succeeded in pushing through Congress new sugar legislation. The Sugar Act of 1937 circumvented the Agricultural Adjustment Act and restored payments to growers to maintain the quota system.[96]

As the Depression continued to take its toll on the nation and the sugar industry, events in Europe and Asia were occurring which would have a major effect on Louisiana sugar production. The worldwide sugar surplus and the corresponding low prices ended with the coming of World War II in 1939. As a result, shortly after the outbreak of hostilities, sugar quotas were eliminated, but by December, 1939, prices stabilized and President Roosevelt reinstated the quota system. The 1940 season was a poor one for Louisiana production, and the harvest fell below assigned levels.

This was the situation when the United States entered the war against the Axis Powers in December, 1941. Now that the nation was at war, sugar producers took on new importance as sugar was proclaimed to be a critical wartime product, and rationing of sugar was introduced for consumers. The Sugar Act was extended to 1944 so that the government could continue to control sugar prices.

Whereas the coming of war should have translated into good economic times for sugar producers, such was not always the case. Increased industrial production required by the war effort reduced the number of agricultural laborers in the sugar region. More importantly, thousands of men were removed from the labor pool to enter military service. The labor situation had become so bad by 1943 that spokespersons for the sugar industry requested the unprecedented use of prisoners of war incarcerated in Louisiana to harvest that year's cane crop. The result was that roughly 5,000 German prisoners of war helped to harvest the 1943 and 1944 cane crops.

By 1944, the economic problems of the growers and millers had been significantly reduced with the result that the quota system was again cancelled. In 1947, two years after war's end, all rationing and quota stipulations on domestically produced sugar were ended.[97]

Although war should never be given credit for any benefits derived from it, the Louisiana sugar industry did receive an important benefit from World War II: an increase in the speed of mechanization for the cultivation and especially the harvesting aspects of the industry. By 1940 mechanization had been introduced to some aspects of the industry. Road improvements, for example, made by state and federal governments during the Depression, allowed many more growers to transport their sugarcane to the mills by truck. By 1938, one of the new co-ops was receiving 80 percent of its cane by truck, 20 percent by draft-animal carts.[98] The development of better and more powerful tractors with fewer mechanical problems brought more mechanization into the sugarcane fields on the eve of World War II.

The total mechanization of the industry, however, began with the end of World War II. In 1940, 3,300 tractors were being used in the cultivation of sugarcane. By the late 1940s this number had increased to 6,500. Mechanization is reflected in the facts that between 1939 and 1947 sugarcane acreage increased by nearly 20 percent while the labor force associated with the sugar industry declined by 40 percent. This sudden decline in the labor force was owing, of course, to the introduction of the mechanical harvester, which, many growers noted, could cut more cane in one day than could 200 people.

Until the late 1930s there had been few attempts to develop a practical mechanical harvester. One reason for this was that labor in the sugar region was so inexpensive that growers did not feel the need to invest in research necessary to develop a mechanical harvester. Another reason, however, was that technology had not yet reached the point where a practical harvester could be developed for Louisiana sugarcane field conditions. Then, toward the end of the 1930s experimental harvesters began to appear in a few fields. By 1939, 16 of these odd-looking contraptions were systematically mowing down rows of cane. The next year there were 25 harvesters in operation. When World War II ended, and probably as a result of the labor shortages brought on by the war, there were 400 mechanical harvesters in use. Of the 298,000 acres of sugarcane harvested during the 1946 season, 63 percent of the cane was cut by mechanical harvesters.[99] The days of hand-cut sugarcane were numbered.

As the 1940s came to a close, Congress decided to continue its regulation of the sugar industry. In 1948, Congress passed another sugar act with sugarcane growers of Louisiana and Florida receiving a combined quota of 500,000 tons. Sugar acts which in the past had been used to benefit the domestic sugar industry began to be used in the 1950s as a means to hold down consumer sugar prices. The result was further consolidation of the industry. Between 1937 and 1959, the number of sugarcane farms in Louisiana decreased from 10,260 to 2,686, and the number of mills declined from 92 to 47. As this occurred, the number of acres grown per farm increased from 28 in 1937 to 101 in 1959. The seasonal grinding capacity of the mills also increased during these years from 57,000 tons to 116,000 tons per mill. Total sugar production increased while at the same time man hours needed to cultivate a ton of sugarcane decreased.[100] The problem not reflected in these figures was that the price of sugar continued to be below the price needed to pay for production, maintenance, and replacement. Growers were barely receiving enough for their sugar to make a profit, but their investment in the industry was too great for many of them to call it quits.

Even though the federal government was using the sugar act to hold down the price of raw sugar, growers and millers continued to seek ways to improve the cultivation and manufacture of sugar. From the late 1940s onward, chemical control of weeds and pests became the norm in the Louisiana sugar region. Herbicides such as 2-4-D, Dalapon, TCA, and Silvex began to be used in controlling the sugar growers nemesis, johnson grass. Insecticides such as Trichogramma, Endrin, Ryania, and Cryolite, when applied to sugarcane helped to control the sugarcane borer, a pest which attacked the sugarcane stalk causing a reduction of sucrose. In 1953, the discovery of Ratoon Stunting Disease was met quickly and effectively by growers who found that the disease could be controlled by using hot water and hot air treatments on seed cane. While the use of chemicals for weed and pest control was a blessing, it was not long before the industry discovered that certain chemical applications were environmentally unsafe. As these determinations were made and the chemicals removed from the market, growers found other, environmentally safe, herbicides and insecticides.

The 1950s was also the era when sugarcane and raw sugar began to be transported through more efficient means. Sugarcane trucks transporting cane from the fields to the mills became more powerful and less prone to mechanical failure. Trailers in which the cane was placed for transport were improved and became capable of hauling greater amounts of cane. On the milling side, manufacturers began shipping sugar in bulk by railroad cars and by barge. The loading of these transport vehicles was now done by a system of mechanical conveyors.[101]

During the 1960s Louisiana sugar interests continued to maximize their efforts to achieve economy while at the same time promoting the industry's development. Manufacturing became more cost effective. Many more, less efficient, mills closed, and those that continued in operation grew in size (capacity to grind more cane) and efficiency.

Since the 1960s the industry has enjoyed, in general, an era of progress. There were the inevitable freezes (especially the disasterous one of December 1989), droughts, and hurricanes (particularly Hurricane Hilda in 1964, Hurricane Betsy in 1965, and Hurricane Andrew in 1992) to dampen growers' enthusiasm. The oil boom lasting from the late 1970s to the mid-1980s saw much cane land converted to use by the oil industry and related services, but with the end of the boom, sugarcane made a significant comeback.

During the 1960s and 1970s mills continued to develop and install labor-saving devices. At the same time the mills developed ways to use energy efficiently. When it became apparent that there was a need for pollution control, millers began installing devices to keep the environment clean. By the 1970s computers had been introduced into mill operations. The core sampler proved to be an outstanding success in determining the amount of sucrose in a load of cane being delivered to a mill, and therefore the amount of sugar that particular load would produce. Growers and millers benefitted by the development of this technology.[102]

Despite all advances in the fields and in the mills, the 1970s proved to be a period of anxiety for people in the sugar industry. Perhaps a major cause of this was the fact that in the congressional session of 1974, the Sugar Act, which had been the stabilizing force of the sugar industry for the previous 30 years, was up for renewal. Unfortunately for the sugar industry, the coalition of growers, millers, and refiners, which had always been there to insure renewal of the act, broke down shortly before legislation to extend the Sugar Act came up for a vote in Congress. As a result, the Sugar Act expired at the end of 1974. By November, the price of a pound of sugar had risen to 64 cents. This was great news for people in the industry who had experienced low prices during the first years of the decade.[103] But, as everyone knew, these astronomical prices could not last long. They fell quickly to their normal levels which dismayed growers and millers who were, like other Americans, victims of the high inflation rate of the late 1970s.

The 1970s, however, were not all bad for sugar people. Growers, millers, and refiners achieved a great victory when, in 1976, the Louisiana Legislature passed a "Right to Work Law." With the passage of this law, the labor union movement in the sugar region was thought to have been dealt a serious blow in its efforts to unionize sugarcane workers. In reality, however, little happened because the union movement had already suffered greatly as a result of the abortive sugarcane harvest strikes of the early 1950s.[104]

In 1990, Congress passed the Food Security Act which, in essence, serves as the sugar act for the industry. Unlike the earlier Sugar Act, however, the Food Security Act comes up for renewal every five years.

Thus, in the twenty years since the end of the Sugar Act and as the industry comes down to completing the second century of commercial sugar production in Louisiana, one can conclude that through "blood, sweat, and tears" growers, millers, and refiners of the state have overcome monumental obstacles and continue to thrive. Certain trends continue: acreage of sugarcane in the state continues to increase, but may have leveled off with the 1995 growing season; the number of farms continue to decline, but the number of growers may be increasing as more and more farm families introduce children to sugarcane culture as a livelihood; centralization among mills continues with 20 plants making sugar for the 1994 season, but centralization will continue, at least for the next few years. One estimate is that there will be 8 to 12 mills in Louisiana as the new century dawns. New technology continues to affect the industry, one of the latest advances being experimentation with a sugarcane harvester combine.[105]

GLIMPSES

OF THE

MODERN LOUISIANA

SUGAR INDUSTRY

A PHOTO ESSAY

Preparing the Land for Planting Sugarcane

Fertilizing Young Sugarcane (above)
Employing a Mechanical Ditching Machine to Make Drains (below)

Today's Plowman

Using all the Latest Technology

A Mechanical One-Row Cane Cutter (above)
A Mechanical Two-Row Cane Cutter (below)

Setting Fire to Cut Cane to Burn Off the Foliage

A Cane Field Being Burned before the Cane is Taken to the Mill

Today's Mechanical Operation for Loading Cane Carts

Cane Carts Ready to go to the Mill

A Tractor Driver Moves Cane to the Mill

Trucks Used for Highway Transportation of Cane

Cane Carts Ready For Unloading by the Device on the Right

"Feeding the Mill" is a Mechanized Operation (above)
Some of the New Technology: A Core Sampler

Two Louisiana Sugar Mills

A Louisiana Sugarcane Grower and His Family

VI

Louisiana, Free Trade, and the Future

In 1993, the Louisiana sugar industry faced a potentially damaging threat to the continued existence of the industry: the unrestricted free trade of sugar. The United States government was once again planning to enter into a treaty which had the potential of destroying not only the Louisiana sugarcane industry but also the country's sugar beet industry. That treaty was the North American Free Trade Agreement, NAFTA, a treaty between Canada, the U. S., and Mexico which creates a free-trade zone throughout North America. Free trade occurs when all tariffs and duties are removed from goods and they flow unimpeded across national boundaries.

The Louisiana sugar industry regarded NAFTA as a threat because sugar was one of the commodities which would be traded freely by terms of the treaty. Opponents of the treaty feared that Mexico, which was not then an exporter of sugar, would suddenly enter into the world market because the country's large sugar supplies, presently used in cola and candy making, would become available when Mexico imported, tariff free, cheap fructose corn sweeteners from the United States.

The draft treaty stipulated that if Mexico becomes a "net exporter" of sugar during two consecutive years between years seven and fifteen of the treaty's life, then Mexico would receive unlimited access to the U. S. sugar market for its surplus production. Net exporter means that the nation is producing more sugar than it is consuming. If this scenario occurred, the Louisiana sugar industry and the country's beet sugar industry could have been severely hurt and even destroyed because of low prices brought on by huge Mexican sugar exports subsidized by the Mexican government.[106]

As opposition to NAFTA solidified in the state's sugar region, Louisiana's congressional delegation moved to effect a compromise on sugar while still being free to support NAFTA which, overall, was regarded as beneficial to the economic well-being of the nation.

Leading the compromise talks with Mexican and U.S. officials was Senator John Breaux. On November 4, 1993, during the height of the debate on NAFTA, Senator Breaux announced that an agreement beneficial to the Louisiana sugar industry had been reached. The new "side letter" agreement would eliminate most of the stipulations in the NAFTA treaty that benefited Mexican sugar to the detriment of domestic sugar. For example, Mexico would not be able to use high fructose corn sweeteners to create an artificial sugar surplus. In addition, the clause that allowed the Mexican sugar industry free access to the U. S. market if Mexico became a net surplus producer was amended. With the side letter in place, Mexico would be allowed exportation of increased quantities of sugar to the United States if Mexico genuinely became a net surplus producer of sugar. In fact, Mexico would be allowed to increase export levels in years 1 to 7 (1994-2000) to 25,000 tons per year, and in years 7 to 14 increase export levels to 250,000 tons per year premised on being a legitimate surplus producer of sugar.

After the announcement of the side agreement, most opposition to the NAFTA treaty was silenced in Louisiana's sugar region. Even the American Sugar Cane League, which had opposed the treaty from the outset announced its support. Since the treaty went into effect on January 1, 1994, the issue of NAFTA has faded into the background. Growers, millers, refiners, and their families, however, have reason to fear for the future. If nothing is done by Congress to adjust the situation, Mexico will have almost unlimited access to the U. S. sugar market in 2009.[107]

The compromise contained in the side letter of agreement in the NAFTA treaty appears to be a victory for Louisiana sugar people. But the fact remains that the history of sugarcane in Louisiana is the story of survival. For two hundred years, whenever the industry has been threatened by nature or by politics, its defenders, the men, women, and children who depend on sugarcane for homes, schools, medical care, etc., have either achieved victory or have reorganized themselves in the face of defeat. They know that just as surely as they will have to deal with the effects of future hurricanes and freezes so also will they have to deal with politics that intrudes into their livelihoods.

This has been the lot of the Louisiana sugarcane industry for much of the 200 years of its existence. As it closes out the second century of operations, the industry finds itself at a level of remarkable achievement, as attested to by the following facts.[108]

In 1994 sugarcane was being grown in 22 Louisiana parishes.[109] By far, the cane king is Iberia Parish which, in 1993, had 61,801 acres planted for seed and harvest. In 1993 the 22 Louisiana parishes cultivated 393,462 acres of sugarcane which produced 1,872,032,164 pounds of sugar. The value of the 1993 crop was nearly $400,000,000. The direct economic value generated (X 2.75) was $1,072,500,000. For an economically distressed state like Louisiana, sugar obviously plays a major role in the state's economy.

There are approximately 20,000 people directly involved in the industry, but the total number of people benefiting from sugar production is much greater, for 80 percent of all sugarcane acreage is leased from non-sugarcane producing land owners.

As the Louisiana industry enters its third century, all the hopes and fears of the last two centuries attest to the fact that our state's sugarcane people are an enduring lot.

Notes

[1]Edwin Adams Davis, *Louisiana: A Narrative History* (Baton Rouge, 1971), 38; Joe Gray Taylor, *Louisiana: A Bicentennial History* (New York, 1984), 5; Bennett H. Wall, ed., *Louisiana: A History* (Arlington Heights, Ill., 1990), 13-17.

[2]Taylor, *Bicentennial History,* 5; Davis,*Narrative History,* 38; Wall, *Louisiana,* 15-17.

[3]Davis, *Narrative History,* 40-49; Wall, *Louisiana: A History,* 19-22.

[4]Davis, *Narrative History,* 50-57; Taylor, *Bicentennial History,* 9; Wall, *Louisiana,* 34-36.

[5] Davis, *Narrative History,* 133.

[6]Center for Louisiana Studies, *Green Fields: Two Hundred Years of Louisiana Sugar,* (Lafayette, La.: Center for Louisiana Studies, 1980), 4-7; Davis, *Narrative History,* 72.

[7]*Green Fields,* 4-7.

[8]Ibid., 7-8.

[9]William J. Eccles, *The Canadian Frontier: 1534-1760* (New York: Holt Rinehart and Winston, 1969), 174-85; Davis, *Narrative History* , 97; Wall, *Louisiana,* 52-54.

[10]Davis, *Narrative History,* 98-103; Wall, *Louisiana,* 57-61.

[11]*Green Fields,* 1; Davis, *Narrative History,* 133-34.

[12]*Green Fields,* 9.

[13]Lewis Cecil Gray, *History of Agriculture in the Southern United States to 1860,* 2 vols. (reprint ed., Gloucester, Mass.: Peter Smith, 1958), 73-74.

[14]*Green Fields,* 3-4.

[15]Joseph Carlyle Sitterson, *Sugar Country: The Cane Sugar Industry in the South, 1753-1950* (Westport, Conn.: Greenwood Press Publishers, 1973), 10; Arthur Chapman Barnes, *The Sugar Cane* (New York: John Wiley and Sons, 1974), 120.

[16]*Green Fields*, 10-22; Gray, *History of Agriculture*, 739-40; Sitterson, *Sugar Country*, 4-5.

[17]Barnes, *The Sugar Cane*, 84, 216.

[18]Sitterson, *Sugar Country*, 11-12.

[19]Davis, *Narrative History*, 157-64; Wall, *Louisiana*, 82-86.

[20]John Solomon Otto, *Southern Frontiers, 1607-1860: The Agricultural Evolution of the Colonial and Antebellum South* (Westport, Conn.: Greenwood Press, 1989), 120.

[21]Ibid., 120-21; Sitterson, *Sugar Country*, 10-11; Noel Deerr, *The History of Sugar* (London: Chapman and Hall, Ltd., 1950), 478.

[22]Davis, *Narrative History*, 177-86.

[23]Gray, *History of Agriculture*, 740-44.

[24]Ibid., 748.

[25]Ibid., 740; Glenn R. Conrad and Gertrude C. Taylor, "Virginians in the Teche Country, *Attakapas Gazette*, 17 (1982): 10.

[26]Sitterson, *Sugar Country*, 24-25; Conrad and Taylor, "Virginians," 5-6.

[27]Gray, *History of Agriculture*, 749.

[28]Sitterson, *Sugar Country*, 20-21.

[29]Ibid., 113-14; Barnes, *The Sugar Cane*, 120, 128-29.

[30]Sitterson, *Sugar Country*, 115-16; Barnes, *The Sugar Cane*, 120, 128-29.

[31]Sitterson, *Sugar Country*, 115-18; Gray, *History of Agriculture*, 749-50.

[32]Sitterson, *Sugar Country*, 133-35; T. B. Thorpe, "Sugar and the Sugar Region of Louisiana," *Harper's New Monthly Magazine*, 7 (November 1853): 760-61.

[33]Thorpe, "Sugar and the Sugar Region," 762-63.

[34]Sitterson, *Sugar Country*, 137-39.

[35]Ibid., 141-44.

[36]Ibid.

[37]Ibid.

[38]Joseph Carlyle Sitterson, "Financing and Marketing the Sugar Crop of the Old South," *The Journal of Southern History,* 10 (May 1944), 194-96.

[39]Ibid., 196-97.

[40]Sitterson, *Sugar Country,* 197.

[41]Sitterson, "Financing and Marketing," 189-93.

[42]Ibid.; Sitterson, *Sugar Country,* 196-204.

[43]Conrad and Taylor," Virginians," passim.

[44]Gray, *History of Agriculure,* 744-48.

[45]Ibid., 741; John Alfred Heitmann, *The Modernization of the Louisiana Sugar Industry, 1830-1910* (Baton Rouge: Louisiana State University Press, 1987), 26-27.

[46]Sitterson, *Sugar Country,* 138, 282.

[47]Ibid., 139.

[48]Gray, *History of Agriculture,* 741-42; Sitterson, *Sugar Country,* 138.

[49]Gray, *History of Agriculture,* 742; Sitterson, *Sugar Country,* 152-53.

[50]Gray, *History of Agriculture,* 742; Sitterson, *Sugar Country,* 147-50.

[51]Gray, *History of Agriculture,* 742; Sitterson, *Sugar Country,* 147.

[52]Gray, *History of Agriculture,* 742; Sitterson, *Sugar Country,*150-52.

[53]Gray, *History of Agriculture,* 742; Deerr, *History of Sugar,* 573-74.

[54]David W. Francis, "Antebellum Agricultural Reform in *DeBow's Review, Louisiana History,* 14:165-66.

[55]Ibid., 172-74.

[56]Ibid., 176-77.

[57]Sitterson, *Sugar Country,* 50, 60; Mark Schmitz, *Economic Analysis of Antebellum Sugar Plantations in Louisiana* (New York: Arno Press, 1977), 108-9; Joe Gray Taylor, *Negro Slavery in Louisiana* (New Orleans: Louisiana Historical Association, 1969), 95-97.

[58]Sitterson, *Sugar Country,* 61.

[59]Ibid., 89-92; Davis, *Narrative History,* 211.

[60]Wall, *Louisiana,* 119-27.

[61]Ibid., 130-31.

[62]Davis, *Narrative History,* 244-45.

[63]Ibid., 245-48.

[64]Charles P. Roland, *Louisiana Sugar Plantations during the American Civil War,* (Leiden, Netherlands: E. J. Brill, 1957), 23-25.

[65]Ibid., 29-36; Davis, *Narrative History,* 254.

[66]Roland, *Louisiana Sugar Plantations,* 30-31, 40-45; Davis, *Narrative History,* 254.

[67]Roland, *Louisiana Sugar Plantations,* 50 57.

[68]Walter Prichard, "The Effects of the Civil War on the Louisiana Sugar Industry," *Journal of Southern History,* 5:319; Roland, *Louisiana Sugar Plantations,* 104-07; Charles P. Roland, "Difficulties of Civil War Sugar Planting in Louisiana," *Louisiana Historical Quarterly,* 38:41-44.

[69]Roland, *Louisiana Sugar Plantations,* 65-71; Roland, "Difficulties," 45-61.

[70]Roland, *Louisiana Sugar Plantations,* 57-60; Davis, *Narrative History,* 256-57.

[71]Prichard, "Effects," 320; Roland, "Difficulties," 53-54.

[72]Davis, *Narrative History,* 257-59; Roland, *Louisiana Sugar Plantations,* 57, 65.

[73]Davis, *Narrative History,* 261, 268-69; Roland, *Louisiana Sugar Plantations,* 112.

[74]Prichard, "Effects," 321-23; Davis, *Narrative History,* 262-64; Roland, *Louisiana Sugar Plantations,* 115.

[75]Wall, *Louisiana,* 197.

[76]Ibid., 198-203.

[77]Ibid., 206-7.

[78]Ibid., 213-18; William Ivy Hair, *Bourbonism and Agrarian Protest: Louisiana Politics, 1887-1900* (Baton Rouge, Louisiana State University Press, 1969), 107, 110-11.

[79]J. Vincenza Scarpaci,"Labor for Louisiana's Sugar Cane Fields: An Experiment in Immigrant Recruitment," *Italian Americana* (Spring-Summer 1981):19; Mark Schmitz, "Transformation of the Southern Cane Sugar Sector: 1965-1930," *Agricultural History,* 53 (January 1979):271-73.

[80]Scarpaci, "Labor," 22-30; Richard J. Amundson, "Oakley Plantation: A Post Civil War Venture in Louisiana Sugar," *Louisiana History,* 9 (1968), 26-31.

[81]Heitman, *Modernization,* 72-78.

[82]Ibid., 107-9, 134.

[83]Ibid., 125, 170-73, 200-2, 218-19.

[84]Ibid., 90-95.

[85]Schmitz, "Transformation," 282; Heitmann, *Modernization,* 217; Donald W. Millet, "Some Aspects of Agricultural Retardation in Southwest Louisiana, 1865-`900," *Louisiana History,* 11 (1970), 53-54.

[86]Schmitz, "Transformation," 283-84; Heitman, *Modernization,* 249; Millet, "Some Aspects," 54; Marshall E. Schott, "Louisiana Sugar and the Cuban Crisis, 1895-1896," *Louisiana History,* 31(1990), 265-72.

[87]Sitterson, *Sugar Country,* 259-64; Millet, "Some Aspects," 56-57; Ralph Shlomowitz, "Plantation Smallholdings: Comparative Perspectives from the World Cotton and Sugar Cane Economies, 1865-1939," *Agricultural History,* 58 (1984):6-8.

[88]Sitterson, *Sugar Country*, 259-64; Heitmann, *Modernization*, 216; Louis Ferleger, "Cutting the Cane: Harvesting in the Louisiana Sugar Industry," *Southern Studies*, 23 (1984):45, 48-49, 55-57.

[89]Heitman, *Modernization*, 102-3, 245-50, 264-65, 267.

[90]Sitterson, *Sugar Country*, 347-49, 352-58.

[91]Ibid.

[92]Ibid., 343-46; 357-58; 379-82, 386; interview with Julian F. Conrad, New Iberia, La., October 16, 1994.

[93]Sitterson, *Sugar Country*, 346-47, 382-83; Schmitz, "Transformation," 274-75, 277-80.

[94]A. B. Gilmore, pub., *The Louisiana Sugar Manual*, 28 ed., 1938-1939, p. 116.

[95]Sitterson, *Sugar Country*, 380-84, 388; *Green Fields*, 83; Thomas Becnel, *Labor, Church, and the Sugar Establishment: Louisiana, 1887-1976* (Baton Rouge: Louisiana State University Press, 1980), 34-35.

[96]Sitterson, *Sugar Country*, 384-85; Becnel, *Labor*, 34-35.

[97]Sitterson, *Sugar Country*, 393-94; Joseph T. Butler, Jr., "Prisoner of War Labor in the Sugar Cane Fields of Lafourche Parish, Louisiana: 1943-44," *Louisiana History*, 14:283-84, 286-87.

[98]Gilmore, *Sugar Manual*, 116.

[99]Sitterson, *Sugar Country*, 386-87, 394-95; Butler, "Prisoner of War Labor," 285-86; Geoff Burrows and Ralph Shlomowitz, "The Lag in the Mechanization of the Sugarcane Harvest: Some Comparative Perspectives," *Agricultural History*, 66 (1992), 64-67, 73.

[100]Sitterson, *Sugar Country*, 393; *Green Fields*, 83-84.

[101]*Green Fields*, 88-93.

[102]Ibid., 99; American Sugar Cane League, *The Louisiana Sugar Industry* (Thibodaux, La.: American Sugar Cane League, 1993), 2.

[103]*Green Fields*, 115, 135-36; Becnel, *Labor*, 220.

[104]Becnel, *Labor*, 199-200.

[105]*Louisiana Sugar Industry*, 2.

[106]Tim Burley, "The Effects of NAFTA (as proposed) on the Louisiana Sugar Industry," *The Sugar Bulletin*, April 25, 1993, 11-13, 15-17.

[107]Don Wallace, "Washington Update: NAFTA Passes House; Sugar Deal Included," *The Sugar Bulletin*, 72 (December 1993), 5, 16-17.

[108]*The Louisiana Sugar Industry*, 2.

[109]Ibid.

BIBLIOGRAPHY

Books

Barnes, A. C. *The Sugarcane.* Aylesbury: Leonard Hill Books, 1974.

Becnel, Thomas A. *Labor, Church, and the Sugar Establisment: Louisiana, 1887-1976.* Baton Rouge: Louisiana State University Press, 1980.

Butler, W. E. *Down Among the Sugar Cane: The Story of Louisiana Sugar Plantations and their Railroads.* Baton Rouge: Moran Publishing Corporation, 1980.

Center for Louisiana Studies. *Green Fields: Two Hundred Years of Louisiana Sugar. A Catalog Complementing the Pictorial Exhibit.* Lafayette, Louisiana: Center for Louisiana Studies, 1980.

Chen, James C. P. *Cane Sugar Handbook.* New York: John Wiley and Sons, 1985.

Davis, Edwin Adams. *Louisiana: A Narrative History.* Baton Rouge: Claitor's Publishing Division, 1971.

Deerr, Noel. *The History of Sugar.* 2 vols. London: Chapan and Hall, Ltd., 1950.

Edson, H. *Sugar from Scarcity to Surplus.* New York: Chemical Publishing Company, 1958.

Galloway, J. H. *The Sugar Cane Industry: An Historical Geography from its Origins to 1914.* New York: Cambridge University Press, 1989.

Gray, Lewis Cecil. *History of Agriculture in the Southern United States to 1860,* 2 vols. Gloucester, Mass.: Peter Smith, 1958.

Hair, William Ivy. *Bourbonism and Agrarian Protest: Louisiana Politics, 1887-1900.* Baton Rouge: Louisiana State University Press, 1969.

Heitmann, John Alfred. *The Modernization of the Louisiana Sugar Industry, 1830-1910.* Baton Rouge: Louisiana State University Press, 1987.

Hess, John L. and Karen Hess. *The Tastes of America.* New York: Grossman Publishers, 1977.

Howell, Roland Boatner. *Louisiana Sugar Plantations, Mardi Gras and Huey P. Long: Reminiscences of Roland Boatner Howell, Sr.* Baton Rouge: Claitor's Publishing Division, 1969.

Mintz, Sidney W. *Sweetness and Power: The Place of Sugar in Modern History.* New York: Elizabeth Sifton Books, Viking, 1985.

Otto, John Solomon. *The Southern Frontiers 1607-1860: The Agricultural Evolution of the Colonial and Antebellum South.* New York: Greenwood Press, 1989.

Roland, Charles P. *Louisiana Sugar Plantations during the American Civil War.* Leiden, Netherlands: E. J. Brill, 1957.

Schmitz, Mark. *Economic Analysis of Antebellum Sugar Plantations in Louisiana.* New York: Arno Press, 1977.

Sehalit, Michael. *Guide to the Literature of the SugarIndustry.* New York: Elsevier Publishing Company, 1970.

Sitterson, J. Carlyle. *Sugar Country: The Cane Sugar Industry in the South, 1753-1950.* Lexington, Kentucky: University of Kentucky Press, 1953.

Spencer, Guilford L., and George P. Meade. *Cane Sugar Handbook.* 10th ed. New York: John Wiley and Sons, 1977.

Taussing, Charles William. *Some Notes on Sugar and Molasses.* New York: n.p., 1940.

Taylor, Joe Gray. *Louisiana: A Bicentennial History.* New York: W. W. Norton & Co., Inc., 1976.

______________. *Negro Slavery In Louisiana.* New Orleans: Louisiana Historical Association, 1969.

Van Hook, Andrew. *Sugar: Its Production, Technology, and Uses. New York; Ronald Press Company, 1949.*

Wall, Bennett H., ed. *Louisiana: A History.* 2nd ed. Arlington Heights, Ill., 1990.

Articles

Amundson, Richard J. "Oakley Plantation: A Post Civil War Venture in Louisiana Sugar." *Louisiana History* 9 (1968): 21-42.

Becnel, Thomas A. "The Ellenders: Pioneer Terrebonne Parish Family, 1840-1924." *Louisiana History,* 26 (1985): 117-28.

Burrows, Geoff, and Ralph Shlomowitz. "The Lag in the Mechanization of the Sugarcane Harvest: Some Compara-tive Perspectives." *Agricultural History* 66 (1992): 61-75.

Butler, Joseph T., Jr. "Prisoner of War Labor in the Sugar Cane Fields of Lafourche Parish, Louisiana: 1943-1944." *Louisiana History* 14 (1973): 283-96.

Clayton, J. E., and B. R. Eliand. "Methods for Moving High Tonnage Biomass from Field to Furnace." *Agricultural Engineering* 65 (1984): 21-22.

Conrad, Glenn R. and Gertrude C. Taylor. "Virginians in the Teche Country," *Attakapas Gazette* 17 (1982): 5-18.

Damann, K. E., and C. A. Hollier. "Distribution and Incidence of Ratoon Stunting Disease in Louisiana Sugar-cane." *Plant Disease* 75 (1991): 566-71.

Ferleger, Louis. "Cutting the Cane: Harvesting in the Louisiana Sugar Industry." *Southern Studies* 23 (1982), 21-34.

____________. "Farm Mechanization in the Southern Sugar Sector after the Civil War," *Louisiana History* 23 (1982), 21-34.

Fox, William Lloyd. "Harvey W. Wiley's Search for American Sugar Self-Sufficiency." *Agricultural History* 54 (1980): 516-26.

Francis, David W. "Antebellum Agricultural Reform in *De Bow's Review," Louisiana History* 14 (1973): 165-78.

Gaiennie, Betty S. "Sugarcane Retreats South," *Louisiana History* 25 (1984): 76-78.

Heitmann, John A. "Organization as Power: The Louisiana Sugar Planters' Association and the Creation of Scientific and Technical Institutions, 1877-1910," *Louisiana History* 27 (1986): 281-94.

______________. "Responding to the Competition: The Louisiana Sugar Planters' Association, the Tariff, and the Formation of the Louisiana Sugar Exchange, 1877-1885," *Southern Studies* 25 (1986): 315-40.

Hilliard, Sam B. "Site Characteristics and Spatial Adaptability of the Louisiana Sugarcane Industry." *Agricultural History* 53 (1979): 254-69.

McGinty, G. W. "Changes in Louisiana Agriculture, 1860-1880," *Louisiana Historical Quarterly* 18 (1935): 407-29.

Millet, Donald W. "Some Aspects of Agricultural Retardation in Southwest Louisiana, 1865-1900," *Louisiana History* 26 (1985): 301-12.

Nordmann, Chris. "A Commitment to Leisure: The Agricul-tural Economy of St. Landry Parish, La., 1850," *Louisiana History* 26 (1985): 301-12.

Prichard, Walter. "The Effect of the Civil War in the Louisiana Sugar Industry," *Journal of Southern History* 5 (1939): 315-32.

Roland, Charles P. "Difficulties of Civil War Sugar Planting in Louisiana," *Louisiana Historical Quarterly* 38 (1955): 40-62.

Schmitz, Mark. "The Transformation of the Southern Cane Sugar Sector, 1860-1930," *Agricultural History* 53 (1979): 270-85.

Schott, Marshall E. "Louisiana Sugar and the Cuban Crisis, 1895-1898," *Louisiana History* 31 (1990): 265-72.

Shlomowitz, Ralph. "'Bound' or 'Free'? Black Labor in Cotton and Sugarcane Farming, 1865-1880," *Journal of Southern History* 50 (1984): 569-96.

_______________. "Plantations and Small Holdings: Comparative Perspectives from the World Cotton and Sugar Cane Economies, 1865-1939," *Agricultural History* 58 (1984): 1-16.

Sitterson, J. Carlyle. "Financing and Marketing the Sugar Crop of the Old South," *Journal of Southern History* 10 (1944): 188-99.

_______________. "The McCollams: A Planter Family of the Old and New South," *Journal of Southern History* 6 (1940): 347-67.

_______________. "The William J. Minor Plantations: A Study in Ante-Bellum Absentee Ownership," *Journal of Southern History* 9 (1943): 59-74.

Thorpe, T. B. "Sugar and the Sugar Region of Louisiana," *Harpers New Monthly Magazine* 7 (1853): 746-67.

Whitten, David O. "Sugar Slavery: A Profitability Model for Slave Investments in the Antebellum Louisiana Sugar Industry," *Southern Studies* 12 (1973): 423-42.